Agricultural Extension - A Glimpse

Dr. S. JANANI
Ph.D,
Assistant Professor
Adhiyamaan College of Agriculture and Research,
Athimugam, Krishnagiri,
Tamil Nadu.

Mr. R. HARISH
Masters in Business Administration,
Assistant Professor
Adhiyamaan College of Agriculture and Research,
Athimugam, Krishnagiri,
Tamil Nadu.

Dr. T.N. SUJEETHA
U.G., P.G., Ph.D
Research Associate
Department of e-extension,
Tamil Nadu Agricultural University,
Coimbatore.

Dr. M. ASOKHAN
Ph.D., Professor and Head,
Agricultural Extension from Tamil Nadu Agricultural University
Coimbatore.

SCIENTIFIC PUBLISHERS

Published by
SCIENTIFIC PUBLISHERS (INDIA)
5 A, New Pali Road, P.O. Box 91
Jodhpur 342 001, INDIA
E-mail: info@scientificpub.com
Website: http://www.scientificpub.com

ISBN: 978-93-88812-87-0
eISBN: 978-93-88812-89-4

Printed in India

Dr. S. JANANI

Dr. S. JANANI has completed her Ph.D degree in Agricultural Extension in Agricultural College and Research Institute, Tamil Nadu Agricultural University, Coimbatore. She has done her doctoral research on "Entrepreneurial Orientation of Rural Youth - An Analysis" and her master's degree research was on "Training Needs of Seed Growers and Seed Dealers with respect to Promoting Quality Pulses Seed Production". She has published several research articles, popular articles and review articles in reputed national and international journals. She has presented several research papers and posters in national and international seminars. She has won "Young Scientist Award". She has been awarded with Rajiv Gandhi National Merit Fellowship for pursuing her doctoral degree. She has authored books titled "Extension Methods at a Glance" and "Objective Agricultural Sciences". Presently she is working as a Assistant Professor (Agricultural Extension) in Adhiyamaan College of Agriculture and Research, Athimugam, Krishnagiri, Tamil Nadu.

Dr. T.N. SUJEETHA

The author is a doctorate in the field of Agricultural Extension. She has completed her U.G., P.G. and Ph.D degree programmes in Agricultural College and Research Institute, Tamil Nadu Agricultural University, Coimbatore, Tamil Nadu. She has published several research articles, popular articles and review articles in reputed national and international journals. She has participated in several national and international conferences and has won several awards viz., Best Poster Award, Best Presenter Award and Best Paper Award in the field of Agricultural Extension. She has been awarded with Rajiv Gandhi National Merit Fellowship for pursuing her doctoral degree. Presently she is doing Post Doctoral Research Fellowship in the department of Agricultural Extension and Rural Sociology, Tamil Nadu Agricultural University, Coimbatore.

Mr. R. HARISH

Mr. R. Harish is Masters in Business Administration, with specialization in Agribusiness from Agricultural College and Research Institute, Tamil Nadu Agricultural University, Coimbatore. He is having a work experience of three and half years in the field of agriculture marketing as Territory Manager and Management Sales Trainee from Monsanto India Limited, a leading agriculture input industry. He has also been involved in R&D as Manager Business Incubation and Agriculture Extension in forming and registration of three Farmers producer companies funded by NABARD. Presently he is working as Assistant Professor (Agriculture and Rural Management) in Adhiyamaan College of Agriculture and Research, Athimugam, Krishnagiri, Tamil Nadu.

Dr. M. ASOKHAN

M. Asokhan is a Professor and Head, Department of Agricultural Extension and Rural Sociology, TNAU, Coimbatore. M. Asokhan was awarded Ph.D. degree in Agricultural Extension from Tamil Nadu Agricultural University Coimbatore, where he started his professional career as Assistant Professor in Agricultural Extension. He has experience of over 18 years in teaching, research & extension. He has served as Programme coordinator KVK and Professor and Head, in Sugarcane Research Station. He has organized several on-campus and off campus trainings. He has published several research papers and articles in various reputed journals, books and magazines. He has acted as Vice-President of Society of Extension Education, TNAU, Tamil Nadu. He has also acted as member in various selection committee. He has guided considerable number of students in the extension discipline and he acted as Advisory Committee Members for students from different disciplines like Rural Management, Economics, Agronomy and other disciplines. He has completed more than eight research projects.

PREFACE

Agricultural Extension is the application of scientific research and knowledge to agricultural practices through farmer education. It plays a crucial role in promoting agricultural productivity, increasing food security, improving rural livelihoods, and promoting agriculture for the economic growth of every country.

The study of Agricultural Extension as a subject is interwoven with different concepts, facts and figures. The objective of preparing this book is to help the readers to quickly grasp the concepts in a comprehensive, systematic and effective manner.

The compilation of this book is mainly to present the relevant concepts being asked in various examinations being conducted by ASRB, ARS, SRF, JRF/NET and other entrance examinations. The concepts under each heading gives an overall idea about different aspects of Agricultural Extension. It gives insights and acts as a quick guide on Agricultural Extension. Moreover, the abbreviations given at the end will give the candidates an overall idea about the questions being asked in competitive examinations.

Since efforts were made to compile the reliable information, there may be some important concepts be missing. If so, the readers are welcome to give their valuable suggestions and to point out errors, if any, for improving the quality of the book.

Authors

CONTENT

1. Rural Sociology and Educational Psychology 3
2. Communication and Extension Teaching Methods 25
3. Farm Journalism 40
4. Diffusion and Adoption of Innovation 44
5. Training and Training Methods 51
6. Participatory Methods and Programme Planning in Extension 58
7. Extension Education, Research Design, Extension Pluralism and Group Led Extension 70
8. e-Extension 81
9. Management and Entrepreneurship 92
10. Abbrevations 104
11. Reference 109

CHAPTER 1

Rural Sociology and Educational Psychology

Rural sociology is the field of sociology that is associated with the scientific study of the rural society. It studies about the culture, customs, social life, social values, social stratification, social institutions, social organizations, social interactions, social groups and so on that are prevailing in the rural society. Educational psychology deals with the study of human behaviour in the educational setting. For an extension worker, the area of operation is village, so an extension worker must have a good knowledge about the people in rural areas.

Accommodation:

Accommodation is a process of developing temporary working agreements between conflicting individuals or groups.

Acculturation:

Acculturation refers to those phenomena that occur when people of different cultures come into continuous contact, resulting in subsequent changes in the original cultural pattern of either one culture or both.

Achieved Social Role:

It is the role which are earned or achieved by the people through individual effort and competition. Eg: Elected leaders.

Adelphic or Fraternal Polyandry:

One wife is regarded as the wife of all brothers.

Affinal Kinship:

The bond of a marriage is called affinal kinship. When a person marries, he establishes relationship not only with the girl whom he marries but also with a number of other people in the girl's family.

Aggregation:

It is the collection of individuals in physical proximity of one another. There may be some interaction between the individuals but it is generally of a temporary nature and lacks definite pattern of organization.

Anomic:

The term is used to describe the state of failure of norms or normlessness. The word is French for 'lack of rules' and refers to social situations in which many norms exist but do not really bind all members of society.

Anti-Social Group:

An anti-social group is one that acts against the interest of society.

Aptitude:

It is an ability to form approaches. It is the ability to form a certain kind of habit efficiently.

Ascribed Social Role:

It is the role which is assigned by culture and society to an individual. Eg: to become head of household.

Assimilation:

It is referred as fusing or blending process whereby cultural differences tend to disappear and individuals and groups once disappear become similar.

Association:

Association is defined as the group of people who are organized for a common interest. It can also be defined as an organization deliberately formed for the collective pursuit of some interest or set of interests, which its members share.

Atomistic Type:

When the individual's rights are maximum and considered supreme, such type is called atomistic type.

Attention:

This is the process of attending to series of stimuli from among the many stimuli which are within range. Psychologically we select only those that are related to our present needs and interests.

Attitude:

An attitude is an emotionalized system of ideas which predisposes an individual to act in a certain way under certain conditions. Much of our learning involves change in our attitudes, our disposition to give favourable or unfavourable response to objects, persons, situations or abstract ideas.

Audience:

Audience is a crowd formed for a specific period of time based on some specific laws.

Behaviour:

It is the expression of one's thoughts and feelings.

Band:

Band is the original form of local group which stands at the lowest level of social organization-that of migratory hunters. It is a small group of people which possesses a strong community feeling. It is generally nomadic and does not have a definite place of settlement.

Belief:

Belief is the mental convictions one has about the truth or actuality of something.

Case Study:

It is a qualitative analysis consisting of careful and complete observation of an individual or a social situation or a social unit. It is an in-depth study.

Caste:

An aggregate of persons whose share of obligations and privileges are fixed by birth, sanctioned and supported by religion and usage.

Category:

Category means collection of items that have atleast one common characteristic that distinguishes from other items which have other characteristics in common.

Clan or Sib:

It is the simplest type of unilateral group which is extended to include all those believed to be related through common descent. e.g., Gotra system in Hinduism. In other words, a clan is the group of individuals who believe themselves to be the common descendents of a real or mythical ancestor.

Class:

It is an abstract category of persons arranged in levels according to the social status they possess. In other words, it is one or more broad groups of individuals who are ranked by the members of the community in socially superior and inferior positions.

Cognition:

The mental activity or the combination of mental activities, resulting in the knowledge or being aware of something.

Collectivity:

It is a number of persons whose behaviour is specifically polarised around a temporary centre of attraction that leads to interaction and a unity which exists as long as centre of attraction exists.

Community:

It refers to the groups belonging to particular area and religious faiths such as Hindhu community, Sikh, Muslim etc. Community may also be defined as the group of people living within a continuous geographic area, sharing common values and a feeling of belonging to the group, who come together in the common concerns of daily life.

Competition:

Competition is the social process or form of social interaction in which two or more individuals or groups strive against each other for the possession or use of some material or non-material good.

Conception of Self:

It is the idea that a person holds about himself or herself and depends largely on the way others act towards him or her.

Conditioned Response Learning:

When the learning takes place based on the identical past experience, such type of learning is called conditioned response learning.

Conflict:

It refers to the struggle in which competing parties, attempting to reach a goal, strive to eliminate an opponent by making the other party ineffectual or by annihilation.

Congregate Group:

It is the voluntary group in which he/she moves or which he/she joins voluntarily.

Conjugal Family:

If the family comprises of husband, wife and adopted children, i.e., no blood relations, then it is called conjugal family.

Connectors:

Connectors are those which connects the receptors and effectors.

Consanguine Family:

Based on blood relationship, consisting of a large group of blood relatives with a fringe of spouses.

Consanguinous Kinship:

The bond of blood is called consanguinous kinship. The consanguinous kin are related through blood. The bond between parents and their children and that between siblings is consanguinous kinship.

Conservatism:

It is the emphasis on the past, preserving it and continuing it.

Cooperation:

Cooperation means working together towards common objectives or goals.

Crowd:

The crowd may be defined as a collection of individuals united temporarily and in close proximity to each other whose object may be of diverse kinds.

Cultural Inertia:

It is the resistance to change by people in a culture.

Cultural Lag:

Cultural lag occurs when some parts of a people's culture do not change as do other parts. It means that one or more phases of culture moves ahead, while others lag behind.

Culture:

Culture is the complex whole which includes the knowledge, belief, art, moral, law, custom and any other capabilities and habits acquired by man as a member of society.

Culture Complex:

A culture complex is a group or cluster of related culture traits.

Culture Pattern:

It is a group of culture complexes, formed in the same way to form a culture pattern.

Culture Trait:

Culture trait is the smallest unit of culture and has been defined more specifically as a reportedly irreducible unit of learned behaviour pattern of material product thereof.

Custom:

Custom is the practices that have been often repeated by a multitude of generations, practices that tend to be followed simply because they have been followed in the past.

Development:

It refers to growth, evolution, stage of inducement or progress. This progress of growth is gradual and has sequential phases.

Ecological Entity:

People in society distribute themselves on a territorial geographic basis and through relationship with each other create land based units, called ecological entity.

Educational Psychology:

Educational psychology is the systematic study of development of individual within the educational settings.

Emotion:

Emotion is a state of being moved, stirred up or aroused and involve impulses, feelings and physical and psychological reactions.

Empathy:

The ability to feel another's feelings in the way the other person does and the ability to communicate that similarity in feeling is termed as empathy.

Endogamous Family:

Affiliation between two in-groups is called endogamous family.

Endogamy:

Marrying with one's tribe or caste. In our country, marriage of Hindus is endogamous.

Environment:

It is anything that surrounds or environs us. Environment in this sense is made of all those things which though distinct from us affect our life or activity in some way or the other.

Effectors:

The parts of body by which man's actions to object and persons are performed are called motor organs or organs of response.

Ethics:

Ethics is the system of moral principles. The branch of philosophy which deals with what is good or bad, and right or wrong, and with moral duties and obligations is called ethics.

Ethnic Group:

Any kind of group, racial or otherwise, which is socially identified as different and has developed its own sub culture is called ethnic group.

Ethnocentrism:

Ethnocentrism is the tendency of man to consider his own culture as high value and superior to all others, and to judge other cultures in terms of standards and values that exist in one's own culture.

Ethos:

Ethos is a term used to refer to the core of norms and values that characterize a sub-culture or culture and give it a 'feeling' or tone.

Exogamous Family:

Affiliation between two out-groups is called exogamous family.

Exogamy:

Marrying outside one's clan because followers of the system believe that there should not be inbreeding or sexual intimacy between primary kins.

Extended or Joint Family:

Joint family is an aggregate of more than one primary family, on the basis of close blood relationships and common residence.

Familism:

It is the subordination of individual accomplishments and goals to those of the family.

Family:

Family is a group of persons, united by ties of marriage, blood or adoption, constituting a single household, interacting and communicating with each other in the respective social role of husband and wife, son and daughter, mother and father, brother and sister, creating and maintaining a common culture.

Fatalism:

It is the belief in fatem, rather than in one's own ability to change things. Fatalism is a feeling that one cannot control events or prevent unpleasant things from happening.

Folkways:

Folkways are the customary ways of behaviour in society in which society exerts some, and only some, force for conformity. Violation of folkways is not taken seriously by the society.

Formal Group:

Formal groups are organized, have a membership role, defined roles, rules or procedures of operation and rigidly enforced behaviour of its members. Example: Panchayats, registered societies etc.

Frustration:

The term is used for the condition in which an individual's wish/goal is blocked or is discovered to be unattainable.

Fundamental Rights:

Fundamental right is a legally enforceable right governing the relations between the state and the individual. These rights are essential to provide security and equality of citizenship and thereby helping the process of nation building.

Gemeinschaft:

The gemeinschaft is a society in which most relationships are traditional or personal or often both.

Genetic Group:

Genetic group is the family in which a man is born involuntarily.

Geographical Social Mobility:

It refers to the movement of a group from one geographic area to another.

Geseilschaft:

Geseilschaft is a society in which there is neither personal attachment nor the importance of traditional rights, obligations and duties.

Group Marriage:

The marriage of a number of men to a number of women is called group marriage.

Group:

A group is an aggregate of two or more people in reciprocal interaction or communication with each other.

Group Norms:

Group norms are the standards of behaviour and performance expected for members.

Hallucination:

Perceiving a figure or an object purely because of our subjective conditions, when there is no stimulus at all. Such an error in perception which has no basis in a real sensory stimulus is called hallucination.

Heredity:

Heredity is a biological process of transmission of certain traits of behaviour of the parents to their children, by means of the fertilized egg.

Homogeneity:

The similarity of social and psychological characteristics in the population as language, beliefs, mores and patterns of behaviour is called homogeneity and the dissimilarity is called heterogeneity. Homogeneity is found much more in rural areas whereas heterogeneity is found in urban areas.

Horizontal Group:

The term horizontal group is used to describe members who are alike in the status or position in the class system of society.

Horizontal Social Mobility:

It is defined as the movement of individuals and groups between positions in society which are more or less of the same social status.

Illusion:

An illusion is a wrong or mistaken perception. The perceptual process always involves an interpretation of the sensory experience in the light of our past experience or present attitude, our organic needs etc.

Individualisation:

Individualisation is that social process which tends to make the individual more or less independent of his group and to create in him a self-consciousness of his own.

Informal Group:

Informal groups are not formally organized and rules and procedures are not rigidly enforced. Eg: Friendship groups, play groups etc.

In-Group:

An in-group is a group to which a person feels he belongs with which he identifies himself strongly.

Institution:

Institution is an organized system of social relationships which embodies certain common values and procedures and meets basic needs of the society.

Intelligence:

Intelligence is the ability of an individual to adjust himself to the conditions that arise in his environment.

Interest:

Interest is a preference for one activity over another. It is any aim or object which stimulates activity towards its attainment.

Interview:

Interview is one of the important methods of collecting data and its objective is to exchange ideas or gather information in which one person asks another person either directly or indirectly.

Involuntary Group:

Involuntary group is one, where an individual does not have any choice for his/her membership. Eg. membership by birth.

Isolation:

Isolation is the absence of communicative interaction or social contact. It is the situation deprived of social contacts.

Joint Family:

Joint family is a group of people who generally live under one roof, who eat food cooked at one hearth; who hold property in common and who participate in common worship and are related to each other as some particular type of kindred.

Kinship:

It refers to a social relationship based upon family relatedness as culturally defined. The bond of blood or marriage which binds people together in a group is called kinship.

Knowledge:

Knowledge may be defined as retained information concerning facts, concepts and relationships.

Laws:

Laws generally grow out of mores. Law is the form of power which is socially accepted and has been legitimized.

Leader:

Leaders are persons who are selected by the people because of their special interest or fitness to work on some phase of the local programmes.

Leadership:

Leadership is an activity in which effort is made to influence people to cooperate in achieving a goal viewed by the group as desirable. The directing, influencing and controlling of others in pursuit of a group goal is also termed as leadership.

Learning:

Acquiring or improving the ability to perform a behavioural pattern through experience and practice is called learning.

Marriage:

Marriage is a socially sanctioned union of one or more man with one or more women with the expectations of that they will play roles of husband and wife.

Matriarchal or Matrilineal Family:

When the property inheritance and reckoning of descent take place along a female line.

Matri-local Family:

It is the type of family where the husband goes to live with the family of his wife and the family line goes with the mother.

Monogamous Family:

It is the marriage of one man to one women.

Monogamy:

If one man marries one women at a time, then it is called monogamy.

Mores:

Mores are the positive actions enforcing what ought to be done.

Mores are the patterns of behaviour considered essential by the society. They are the socially acceptable ways of behaviour, that do not involve moral standards.

Motivation:

Motivation is the internal state which stimulates a person to carry out certain activities. Motivation involves arousing, sustaining and directing desirable behaviour.

Motive:

A motive is an energizing condition of an organism that serves to direct that organism towards a certain goal.

Motor or Skill Learning:

It involves primarily the use of muscles of the body. In this, the individual learns muscular coordination as a mode of response to some situation.

Myth:

Myth is an ill-founded belief held uncritically by the people and can be removed through their acquisition of appropriate knowledge.

Neighbourhood:

It can be defined as a limited geographic area in which the individuals and families are known to each other and carry on intimate associations together. It is a smaller unit than the community. A community is often composed of several neighbourhoods.

Non Fraternal Polyandry:

If one women has many husbands with whom she cohabits in turn is called non fraternal polyandry.

Norms:

Norms are the rules that govern action directed towards achieving values. They are the established behaviour patterns for the members of a social system. Norms are the shared conviction about the pattern of behaviour.

Nuclear Family:

Nuclear family or primary family consists of husband, wife and their unmarried children. It is also called as conjugal family.

Opinion Leader:

A person who has a relatively large influence on the opinions of others in the group to which he or she belongs. Opinion leaders are seen as important contributors to the formation of public opinion about new ideas, situation's etc.

Organization:

Organization is those classes of human relationship structure where people purposefully associate in systematically arranged units to promote and achieve some common purposes or interests that are not specifically expressed in the institution's specific roles.

Out-Group:

It is defined in contrast between 'they' and 'we', like we are democrats and they are communists. Eg. We are Hindus and they are Muslims

Patriarchal or Patrilineal Family:

It is the one where determination of descent and property inheritance take place along the male line.

Patri-local Family:

It is the one where the married couple and their offspring put up with the husband's family or in a new household set up by the husband.

Perception:

Perception is the act of interpreting a stimulus registered in the brain by one or more sense of mechanism.

Perceptual Learning:

Based on the past experience people perceive the situation differently. One way of changing a person's habit of responding is to change the way in which the individual perceives the environment.

Personality:

Personality is the sum total of one's experience, thoughts, actions, all behavioural patterns, traits and characteristics that make up a person.

Personality Trait:

Any one aspect or characteristic of a person is referred to as personality trait.

Phratroy:

When a group of clans get together for some purpose, such group is termed as phratroy, e.g., tribal group.

Physiological Needs:

These are the basic needs of the human body such as nourishment or maintenance of body temperature.

Polyandry:

If one women marries several men at a time, then it is called as polyandry.

Polygamous Family:

It if of two types. They are Polygamy and Polyandry

Polygamy Family:

It is a type of family which comprises of marriage of one man with more than one women. A man has more than one wife at a time.

Potential Group:

It is a group made up of number of people having some characteristics common but does not possess any recognizable structure.

Prejudice:

An attitude that predisposes a person to think, perceive, feel and act in favourable or unfavourable ways toward a group or its individual members is called prejudice. Prejudice means to prejudge.

Primary Group:

Primary groups are characterized by intimate face-to-face associations and informal and personal relationships, such as in the family and in groups of close friends.

Pressure Group:

A pressure group is generally an association of persons with a common economic interest who try to influence governmental action in legislation, administrative procedures or judicial decisions.

Pro-Social Group:

Pro-social group is the reverse of anti-social group. It works for the larger interests of the society. It is engaged in constructive tasks and concerned with increasing the welfare of all the people.

Pseudo-Social Group:

A pseudo-social group participates in the larger social life but mainly for its own gain and not for the greater good.

Psychology:

It may be defined as the science of human and animal behaviour.

Public:

Public is a large section of the people who share some common opinion, desire or interest, but who are not organized and are scattered all over. It is an integration of many people not based on personal interaction but on reaction to the same stimuli, a reaction arising without the members of the public necessarily being physically near to one another.

Public Opinion Poll Method:

This method is used to seek and gauge the beliefs, sentiments and attitudes of the public on any given proposition. The results of the public poll help the authorities concerned in modifying their policies accordingly.

Questionnaire:

Questionnaire is a common and popular research tool used to collect data, particularly when data are to be collected from a large number of people who are scattered over a wide area.

Receptors:

It refers to sense organs possessed by organism through which stimuli are received.

Reference Group:

Reference group is a group with which the individual feels identified, the norms and objectives of which he accepts. It is the individual or groups whose behaviour is imitated by him.

Religion:

Religion is a system of beliefs and symbolic practices and objects governed by faith rather than by knowledge, which relates man to an unseen supernatural realm beyond the known and beyond the controllable.

Response:

A response is any resultant activity that is aroused by a stimulus.

Revolution:

Revolutionary movements or revolutions seek to overthrow the existing social system itself and replace it with a greatly different one.

Rules:

Rules are the rights, duties, obligation of any group member which is well defined.

Sanctions:

Sanctions are the means of controlling human behaviour. They may take the form of both rewards and punishments.

Secondary Group:

A secondary group is characterized by few ties of sentiment, formality of contacts and impersonal relationships, such as, may exist between farmer and the commission agent who sells his produce or the wholesale merchant who buys his produce.

Social Change:

Social change is defined as the process for which alteration occurs in the structure and function of a social system.

Social Control:

It can be defined as an application of sanctions to ensure that the members will abide by the group's norms, perform required roles in a prescribed manner and co-ordinate their activities in such a way that group goals can be achieved.

Social Disorganization:

Often, changes in the normal routine life creates a disturbances in the functioning of the existing parts and the customary and socially approved ways of life. Such a situation or state of affairs in society is called social disorganization.

Social Group:

A social group may be defined as two or more persons who are in communication over an appreciable period of time and who act in accordance with a common function or purpose.

Social Institution:

Social institutions are formal, cultural structures devised to meet basic social needs.

Social Interaction:

Social interaction is dynamic interplay of forces in which contact between persons and groups results in a modifications of the attitude and behavior of the participant.

Social Mobility:

Social mobility refers to the movement of individuals and group from one stratum or society to another.

Social Movement:

Social movement is defined as a voluntary association of people engaged in concerted efforts to change attitudes, behaviour and social relationships in a larger society. It is the effort by an association to bring about a change in the society.

Social Processes:

Social processes are the interactions of groups and individuals with one another, and these may take four basic forms:

a) cooperation, b) competition, c) conflict, d) accommodation and e) assimilation.

Social Power:

It is the ability to influence the behaviour of others in accordance with one's ends.

Social Role:

Social role is the expected behaviour of one member of society in relation to others. A single person in society may play a number of roles such as father, teacher, citizen, etc.

Social Security:

Social security is the security that society furnishes through appropriate organization against certain risks to which its members are exposed.

Social Stratification:

Social stratification is the division of a population into two or more layers each of which is relatively homogeneous and between which there are differences in privileges, restrictions, rewards and obligations.

Social Structure:

It is the particular arrangement of the inter-related institutions, agencies and social patterns as well as the statuses and roles which each person assumes in the group. In other words, it is the rural society or interrelationship and interactions of society that provides opportunities for the human being to create forms of association through which society operates.

Social System:

A social system is defined as a set of interrelated units that are engaged in joint problem solving to accomplish a common goal.

Social Values:

Social values are an integral part of culture of society. These are desirable for an organised social life. These are the general conceptions of the good, ideas about the kinds of ends that people should pursue throughout their lives and many different activities in which they engage.

Socialization:

Socialization is a process through which an individual acquires those ways of thinking, feeling and acting which is essential for effective participation within a society.

Society:

Society is defined as the group of people who have lived together long enough, sharing common values and general interests to be considered as a social unit. Society is the web of social relationship.

Sociology:

Sociology is the science of human relationships.

Sociometry:

It is the method which involves identification of leaders and leadership pattern by asking some questions to all persons in the group and tabulating and diagramming their responses.

Spatial Group:

The spatial groups are those which are constituted because of spatial contiguity of their members. Such groups are the clan or sib, tribe and band.

Specialties:

Specialties are elements of a culture shared by some groups but not by all groups in a society.

Statistical Method:

Statistical method refers to the method that is used to measure social problems mathematically.

Status:

Status is a term used to designate the comparative amounts of prestige, difference, or respect accorded to persons who have been assigned different roles in a group or community.

Stereotype:

Stereotypes are the fixed images formed in one's mind about people, practices or various other social phenomena on the basis of experience, attitudes, values, impressions or without any direct experience.

Stimulus:

A stimulus is anything that arouse the organism or any of its parts to activity.

Taboos:

Taboos are the negative actions envisaging what out not to be done.

Trait:

Trait is defined as a typical pattern and resultant behaviour characteristics of a person in a variety of situations.

Tribe:

A tribe is a collection of families bearing a common name, speaking a common dialect, occupying or professing to occupy a common territory, common religious and common culture.

Trustee Type:

The role and behaviour of every individual for the betterment of whole group is called trustee type.

Undefined Acts:

Undefined acts are certain personal habits and patterns of behaviour that have no social approval.

Unique Experiences:

An individual has personal experiences which are unique to him and which deeply influence his behaviour. Such experiences are called unique experiences.

Unsocial Group:

An unsocial group is one which largely lives to itself and for itself and does not participate in the larger society of which it is a part. It does not mix up with other groups and remains aloof from them.

Values:

Values are the conception of the desirable, as standards of evaluation, as guides for decision making behaviour, or simply as expression of preferences. Values are what people consider as valuable and desirable.

Verbal Learning:

The ability to manipulate symbols, as in languages, makes it possible for learning things is called verbal learning.

Vertical Group:

Vertical groups are those groups that are composed of members from different social strata and whose membership cuts vertically across the 'horizontal' groupings in society.

Vertical Social Mobility:

It refers to the two-way movement - up or down of individuals and groups from one stratum in society to another.

Voluntary Group:

Voluntary group is the one where one has to have some deliberate choice for his/her membership. Eg. Cooperative society, women's club etc.

Wishes:

Wish is a pattern of behaviour which involves anticipated future satisfaction, which the person believes is reasonably likely of attainment and towards which the individual usually relates some of his/ her present behaviour.

CHAPTER 2

Communication and Extension Teaching Methods

According to Rogers and Shoemaker (1971), Communication is the process by which messages are transferred from a source to receiver. Van Den Ban and Hawkins (1988) defined commication as the process of sending and receiving messages through channels which establishes common meanings between a source and a receiver. Communication is an art of transmitting information, ideas and attitudes from one person to another. Communication occupies a central position in transfer of technology.

Adaptive or Minikit Trial:

It is a method by which the suitability of a new practice to a given locality under farmer's conditions is determined.

Advertising:

Advertising refers to marketing a product or service in a persuasive manner that encourages the audience to buy the product or use the service.

Amplifier:

It is a device or equipment that makes sound louder.

Audience:

Audience is the consumer of the message. They are the intended receiver of the message.

Audience Response:

Response of the audience is the ultimate objective of any communication function. Response of audience to messages received may be in the form of some kind of action, either mental or physical.

Audience Segmentation:

Audience segmentation is a communication strategy that consists of identifying certain sub audience within a total audience and then conveying a special message to each of these sub audiences.

Audio Aid:

An audio aid is an instructional device in which the message can be heard but not seen.

Audio Teleconferencing:

A telephone or radio network that may be used to provide audio communication among groups at two or more locations is called audio teleconferencing.

Audio-Visual Aid:

An audio-visual aid is an instructional device in which the message can be heard as well as seen. The audio-visual aids include drama, shows, movies, etc. They involve the use of both sound as well as vision, that is, the instructional device through which the message can be heard and seen simultaneously.

Bandwidth:

Bandwidth is a range of frequencies involved in radio and telecommunication transmissions.

Black Board / Chalk Board:

Black board is the most universalized of all the teaching aids. It is not itself a visual material but a vehicle for a variety of visual materials. It is one of the cheapest, most effective, most versatile and easiest to use.

Bioscope:

This is a popular folk medium used for entertainment and for propagation of information on education, agriculture etc. This is helpful in conveying the people about educational messages like vegetable preservation, mosquito control, cultivation of different crops etc.

Broadcasting:

A transmission of programs (radio, television, web based etc.) for public purposes and utilization is called broadcasting. Broadcasting is the act of transmitting audio or visual content through a communication medium, such as radio, television, or film.

Bulletin board:

Bulletin board is a board for display of message. It is a simple inexpensive device that can be placed either outdoor or indoors. A soft board that will hold pins or tacks is most suitable. It can perform basic communication functions. It can attract attention, stimulate interest, deliver a message and promote action.

Bulletin:

It is a printed, bounded booklet with a number of pages, containing comprehensive information about a topic.

Cable Television:

A television service via cable is cable television.

Cable:

Cable is a wire or bundle of wires that carry electric current or a bundle of glass fibers that carry pulses of light. Cables provide the most practical means of transmitting communication signals.

Cablecast:

Any broadcast that is transmitted via a cable television network is called cablecast.

Campaign:

A campaign is an intense educational activity for motivating and mobilizing a community to action, to solve a problem or satisfy a need urgently felt by it.

Cell Phone:

Cell phone is a sophisticated multipurpose telephone which contains a circuit board (the brain of the phone), an antenna, a liquid crystal display (LCD), a keyboard (like a remote control), a microphone, a speaker and a battery.

Chalk board:

Chalk board is a board for writing with chalk which is simplest, cheapest, most convenient and widely used visual aid.

Channel:

Channels are the physical bridges between the sender and the receiver of messages. The channel serves as an essential tool of the communicator.

Charts:

Charts are the visual symbols for summarizing, comparing, contrasting or performing other services, in explaining subject matter. It is the diagrammatic presentation of facts or ideas.

Circular letters:

It is a letter written and sent to many people by the extension worker to publicize an extension activity or to give timely information on farm and home problems.

Code:

Code is a system of signals for communication.

Communication:

Communication is an act by which two or more people share knowledge, feelings, ideas, information etc. in ways such that each gains a common understanding of the meaning, intent and use of the message.

Communicator:

Communicator is a person who provides initiative to the process of communication. He is the source or originator of messages.

Communication Fidelity:

Fidelity is the faithful performance of communication process by all its elements viz., communicator, message, channel and receiver.

Communication Gap:

Communication gap refers to the difference between what was communicated by the extension agent and what has actually been received by the audience.

Communication Network:

A communication network consists of interconnected individuals who are linked by patterned flows of information.

Communication Process:

Communication process is the process that takes place when people or groups communicate with each other.

Credibility:

Credibility is defined as the trustworthiness and competence.

Decoding:

Decoding refers to converting the messages in the code into ordinary language which may be easily understood.

Desktop Videoconferencing:

Desktop videoconferencing is a method for individuals to communicate between single user systems which is growing and changing rapidly. Desktop video conferencing involves two way voice and video contact, but may also encompass other activities such as screen sharing, media display, text based chatting and remote collaboration.

Diagrams:

A diagram is a simplified and structured visual representation of concepts, ideas, constructions, relations, statistical data, etc., used to visualize and clarify the topic.

Diagram, Map and Graph:

Diagram, map and graph are visuals where information is summarized and presented in a more or less abstract form.

Dialogue:

A discussion between an extension agent and a single client to help the client to define a problem and to find a satisfactory solutions.

Diorama:

Diorama is a scenic representation of the original, with specimen, model and painting.

Dash Card and Dip Chart:

These are visual aids in which the messages are written or drawn on thick paper and presented step-by-step by the extension worker to the audience to synchronize with the talk. This produces a dramatic effect on the audience.

Downward Communication:

It has great force for controlling, influencing and indicating activities of organization members.

Drama:

Drama is a theatrical performance around a theme by some people who have rehearsed for it.

Encoding:

Encoding refers to putting the message in code. It is the transformation of messages into signals.

Exhibition:

It is a systematic display of models, specimens, charts, posters and video films in a sequence so as to create interest in the participating farmers.

Extension Agent/ Officer:

A person whose main task is to give extension assistance or to manage an extension organization at the field level is extension agent or extension officer.

Extension Methods:

The methods of communication which can be used in extension for influencing the target groups is called extension methods.

Farm and Home Visits:

It is a direct, face to face contact by the extension worker with the farmer or the members of his family at his home or on his farm for specific purpose.

Farm Clinic:

Farm clinic is a facility developed and extended to the farmers for diagnosis and treatment of farm problems and to provide some specialist advice to individual farmers.

Farm Publication:

It is a class of publications prepared by the extension agency in written form to communicate information. Farm publications are of various types such as leaflets, folder, bulletin, newsletters, pamphlet, extension journals etc.

Farm School on the AIR:

Farm school on the air is a method of providing systematic education on farming to the farmers through the process of distance learning.

Farmer's Call:

Farmer's call is a call made by a farmer or homemaker at the working place of the extension agent for obtaining information and assistance.

Feedback:

Feedback means carrying some significant responses of the audience back to the communicator.

Feedback Loop:

The feedback loop describes the stages that information moves through in an organization from its initial generation to the implementation of changes in products and services.

Field Day or Farmers Day:

It is a method of motivating the people to adopt a new practice by showing what has actually been achieved by applying the practice under field conditions.

Film:

Film is a sequence of images of moving objects photographed by a camera and providing the optical illusion of continuous movement when projected onto a screen.

Film Strips:

Film strip is a continuous strip of film of a small number of photographs, drawings, diagrams arranged in a sequence. A strip may be of 35 mm film.

Flannelgraph:

A flannelgraph or khaddargraph is a visual teaching aid. The pieces of flannel, felt or sand paper, have rough surfaces, will stick to another piece of flannel stretched on a firm flat surface called a 'flannelboard'. When pieces of flannel, or sand paper are attached to the back of the pictures, photographs, drawings, letters etc. these objects will stick to the flannel board. This device is called a flannel graph.

Flash cards:

Flash cards are a series of illustrated cards, which, when flashed or presented to a group, in proper sequence, tells a complete story.

Flip Chart:

Flip chart consist of a series of individual charts which are tagged or bound together and hung on supporting stand.

Folders:

It is a single printed sheet of paper of big size, folded once or twice, and gives essential information relating to a particular topic. It is printed as and when required. It is generally distributed free-of-cost.

Folk Media:

Folk media is a communication system embedded in the culture which existed before the arrival of mass media, and still exists as a vital mode of communication in many parts of the world, presenting a certain degree of continuity, despite changes. It is the communication media which do not use modern technology, but are based on traditional communication methods, such as itinerant stingers, drama or puppet shows.

Gramsat:

Gramsat is tailor-made to meet the basic requirements of rural areas for disseminating region specific knowledge on health, hygiene, environment, family planning and better agricultural practices.

Graphs/Charts:

A graph or chart is a type of graphic information that represents tabular and numeric data. Charts are often used to make it easier to understand large quantities of data and the relationships between different parts of the data.

Group Contact:

In group contact method, the extension agent communicates with the people in groups and not as individual persons. Eg., group meeting, small group meetings.

Group Discussion:

The lecture method may be made more participative, if at the end the audience are allowed to discuss the topic in presence of the speaker and elicit the latter's comments and clarification on the points raised. Learning is reinforced through the interaction of the audience with the speaker.

Horizontal Communication:

It is the flow of information between persons of the same hierarchical level. Horizontal communication is also called as lateral communication.

Illustrations:

Illustrations are non-photographic reconstruction of reality. Eg., drawings, painting, sketches, etc. These are used to illustrate the story or the point the teacher, extension worker makes.

Individual Contact:

The extension agent communicates with the people individually, maintaining separate identity of each person. Eg., Farm and home visit.

Information:

Information is the pattern, people impose on the phenomena they are able to observe.

Interpersonal Communication:

Interpersonal communication refers to that situation where the source and receiver can see each other, talk back and forth, interrupt each other and make responses which both can readily observe.

Journal and Magazines:

These are periodicals, containing information related to various topics of interest.

Keertana or Harikatha:

This is a religious folk theatre. It is a kind of concentrated drama of monodrama in which a gifted actor enters a whole series of characters. Harikatha is also used by central and state governments to educate the masses on family planning, developmental activities, democratic values and national integration with the help of Keertanakar or Kathakars.

Large Group Videoconferencing:

To participate in a videoconference where one or more sites have a large group of people, many logistical factors need to be considered.

Leaflets:

A leaflet is a single printed sheet of paper sometimes folded. It contains accurate or specific information on a specific topic.

Lecture:

This is a formal, verbal presentation of a topic by a single speaker to group of listeners, which may be followed by a question-answer period.

Map:

A map is a simplified depiction of a space, a navigational aid which highlights relations between objects within that space. Usually, a map is a two-dimensional geometrically accurate representation of a three-dimensional space.

Mass Communication:

Mass communication is the process of communicating the proven technologies, information's and ideas to the mass (in large numbers). Mass communication is also called as the channel communication. It is the public communication transmitted electronically or mechanically. In this way messages are transmitted or sent to large, perhaps millions or billions of people spread across the world.

Mass Contact:

The extension agent communicates with a mass of people, without taking into consideration their individual or group identity.

Melas and Festivals:

Melas or Country fairs are synonym with joy in the rural areas where life follows a hard routine, nothing is more welcome to the people than the prospect of a festival and mela. The main objective of these melas is to disseminate information on agricultural technology to the farmers for their betterment.

Message or Content:

A message is an information, a communicator wishes his audience to receive, understand, accept and act upon it.

Method Demonstration:

It is concerned with showing 'how' to do something in a better and easier way.

Method:

A method is a procedure for attaining an objective.

Model:

Models are recognizable imitations or replica of the original, whether workable or not, whether differing or not from the original in size. Models serve to illustrate, to explain and to predict. E.g., farm machinery, models of compost pit etc.

Motion Picture / Cinema:

It is the projected image of a positive print of a sound film taken by a movie camera, which produces synchronized movement of picture with sound.

Nautanki:

Nautanki is one of the most popular folk operatic theatre performance, traditions of South Asia, particularly in Northern India. Nautanki was the single most popular form of entertainment in the villages and towns of Northern India.

Newsletter:

It is a miniature newspaper in good quality paper, containing information relating to the activities and achievements of the organization. It has the fixed periodicity of publication.

Newspaper:

Newspaper is a bunch of loose printed papers properly folded, which contains news, views, advertisements etc. and is offered for sale at regular intervals, particularly daily or weekly.

Non-Projected Audio Visual Aids:

Drama and puppet shows are the most commonly used non-projected audio visual aids.

Non-Projected Visual Aids:

The non-projected visual aids include black board, pictures, posters, charts, flash cards, flip books, flip charts, flannel boards, picture strips, etc. They do not require any equipment, like a projector, for their display.

Non-Interpersonal Communication:

Non-interpersonal communication refers to that particular situation where something (it may be space, time or physical barrier of some sort) is placed between two interacting individuals.

Office Calls:

It is a call made by a farmer or a group at the working place of the extension personnel for obtaining information and assistance.

Online Communication:

Online communication refers to reading, writing and communication via networked computers. It encompasses synchronous computer mediated communication (whereby people communicate in real time via chat or discussion software, with all participants at their computers at the same time); asynchronous computer-mediated communication (whereby people communicate in a delayed fashion by computer, using programs such as e-mail); and the reading and writing of online documents via the World Wide Web.

Opaque Projection:

Opaque projection is a large apertured projector which is used to project opaque materials.

Overhead Projection:

If the picture is projected over the head of the speaker on the screen, such type of projection is called the overhead projection.

Personal Letters:

Personal letter is written by the extension worker to a farmer in connection with extension work.

Photographs:

Photographs are exact visual recordings of things in nature.

Picture:

A picture is a representation made by drawing, painting or photography which gives accurate idea of an object.

Poster:

Poster is a placard displayed in a public place with the purpose of creating awareness amongst the people. It is a visual aid which has been largely used by extension workers in training situation to create awareness, giving information and create interest among the villagers on subjects.

Projected Audio Visual Aids:

Motion pictures or the sound films are the most commonly used projected audio-visual aids.

Propaganda:

A conscious, systematic and organized effort to manipulate the decisions, actions, and ideas of a large number of people, or to influence them with regard to a controversial subject, in a direction which is desirable from the point of view of the insights or interests of the propaganda.

Public Address System:

It is a set of equipments to amplify sound so that it is audible to a large audience over a distance.

Public Communication:

Public communication means communication that is not private to a certain group.

Public Relations:

Public relations is the process of providing information to the public in order to present a specific view of a product or organization.

Puppet Show:

In puppet show, the dolls are made to perform by controlled movement of the puppeteer from outside.

Puppetry:

Puppetry is one of the important traditional media for communicating technology to the farmers in the villages. The art of puppetry was popular both as pure entertainment and as an educational medium. Shadow puppet is a temple ritual performed every year during a temple festival for a specified duration.

Radio Rural Forum:

Radio rural forum is a combination of mass media and inter-personal communication in reaching people with new ideas and persuade them to adopt innovation than mass media alone.

Radio:

Radio is a medium for mass communication, a tool for giving information and entertainment. It is an electronic audio-medium for broadcasting programmes to the audience.

Result Demonstration:

It is concerned with 'why' a practice or input should be adopted by showing the value of new practice over the old practice or of a new practice.

Riddles:

This is also an educational device through which elders used to communicate knowledge. Some puzzles are given to the rural people which help them to understand the use of proper practices in the crop cultivation, home making etc.

Slides:

A slide is a transparent mounted picture which is projected by focusing light through it from electric bulb, petromax or lantern. The projection may be made on roll-back screen or on white wall.

Specimens:

Specimens are real objects taken out of their natural setting. E.g., specimen of insects, ear heads, diseased parts etc.

Story Telling:

Story telling has been one of the best and most commonly used method of instruction in informal education, religious propagation, rural development etc. The listeners make some kind of sound at regular intervals to express attention. The important topics which help in proper development of the rural people can be touched through the stories.

Street Play:

In this form, one narrates a story which is related to a particular theme. These plays are not like theatre but they attach a large number of people. Street plays are increasingly used to spread messages of family planning, sanitation, adult education etc. and they create awareness against social evils.

Study Tour:

A group of interested persons accompanied and guided by one or more extension agents moves out of their neighbourhood to study and learn significant improvements in farm and home elsewhere. The main purpose is to motivate the visitors by showing what others have been able to achieve.

Symbol:

A symbol is a representation of a concept or quantity; i.e., an idea, object, concept, quality, etc.

Talking Doll:

Talking doll is an innovation and may be used to create general awareness of a group of people about an extension programme.

Tamasha:

Tamasha is an extremely lively and best form of folk theatre. It is a traditional Marathi folk art form, often with singing and dancing, widely performed by local or travelling theatre groups within the state of Maharastra.

Tape Recorder:

A tape recorder is an equipment for recording sound on magnetic tape by electro-magnetic process, which may be played back when needed.

Telephone:

Telephone is a system or equipment through which people can communicate both ways to distant places.

Television:

Television is an electronic audio-visual medium which provides pictures with synchronized sound. It is one of the mass media for dissemination of information and technologies in rural areas. It is a system for transmitting visual images and sound that are reproduced on screens, chiefly used to broadcast programs for entertainment.

Translide:

Translides are transparent big size photographs which are displayed by providing light at the back. These are used in communication centre, information centre and exhibitions.

Treatment of Messages:

Treatment means the way a message is handled, dealt with, so that the information gets across to the audience. The purpose of treatment is to make the message clear, understanding and realistic to the audience. It can also be defined as the way a message is processed so that the information gets across to the audience.

Upward Communication:

Upward communication flow is in a direction directly opposite to the flow of official authority, from the dependent subordinates to the superiors or whom they (subordinates) are directly or indirectly dependent for the satisfaction for their needs.

Video:

Video is an electronic audio-video system for preparing video tapes of programmes and events, which could be projected through a receiving set with moving images and synchronized sound.

Video Conferencing:

Video conferencing is conducting a conference between two or more participants at different sites by using computer networks to transmit audio and video data.

Video Teleconferencing:

A video channel added to an audio link between groups by means of satellite transmission, microwave transmission or a two-way cable television system is called video teleconferencing.

Visual aid:

A visual aid is an instructional or communicating device in which the message can be seen but not heard. Visuals help one to communicate effectively.

CHAPTER 3

Farm Journalism

Journalism is defined as the production and distribution of reports on recent events. The main purpose of farm journalism is to provide the farmers with the information they need to make the best possible decisions about their lives, their communities, their societies and their government.

Beats:

Beat reporters are stationed in places where news happens regularly. These strategic locations include police, fire stations, courts, PRO office etc.

Cable Television:

Cable television is a system of distributing television programs to subscribers via radio frequency (RF) signals transmitted through coaxial cables or light pulses through fiber-optic cables.

Community Radio:

Community radio is a type of radio service that caters to the interests of a certain area, broadcasting content that is popular to a local audience but which may often be overlooked by commercial or mass media broadcasters.

Development Journalism:

The term development journalism is used to refer to two different types of journalism. The first idea behind it is similar to investigative reporting, but it focuses on conditions in developing nations and ways to improve them. The other type involves heavy influence from the government of the nation involved.

Etching:

Etching is the process of using strong acid or mordant to cut into the unprotected parts of a metal surface to create a design in intaglio in the metal (the original process - in modern manufacturing other chemicals may be used on other types of material). As an intaglio method of printmaking, it is, along with engraving, the most important technique for old master prints, and remains in wide use today.

Farm Journalism:

Rural/agricultural/farm journalism is a fast moving and a challenging profession to meet the rural/farm information requirements of farm graduates, extension workers, media persons and farmers.

Feature Story:

A feature story is a creative, sometimes subjective article designed primarily to entertain and to inform readers of an event, a situation or an aspect of life.

Folk Media:

Folk media or local media or traditional communication system refers to all organized processes of production and exchange of information. The folk media is a communication vehicle for promoting and improving dialogue which the common people for rural farmers employ to deliver their messages.

Funnel Theory of News Story:

This refers to the technique of journalism which places the most important particulars at the beginning and works 'downstairs' from there.

Hard News:

Hard news is defined as an event that takes place without any notice. It has the element of consequences.

Journalism:

Journalism is the systematic process of gathering, writing, interpreting, processing and disseminating public opinion, public information and public entertainment for publication in newspaper, magazines and broadcast.

Lead:

The lead is the beginning. It is the 'come-on' what the lead says almost always determines whether the reader or listener/viewer will continue with the story or to skip something else.

Libel:

Libel is any false or malicious writing or picture that holds up a person to public ridicule, hatred or contempt or causes injury to his reputation.

Line Art or Line Drawing:

Line art or Line drawing is any image that consists of distinct straight and curved lines placed against a (usually plain) background, without gradations in shade (darkness) or hue (colour) to represent two-dimensional or three-dimensional objects.

News Articles (News Stories):

News is any timely information that interests a number of persons, and the best news is that which has the greatest interest for the greatest number.

News Story:

News story is a piece of information which is enlightening, providing satisfactory details so that the reader gain knowledge of and comprehend what had happened.

News:

News is fresh information concerning something that has recently taken place.

Participatory Journalism:

Participatory or citizen or public or grassroot journalism is the act of people playing an active role in the process of collecting, reporting, analysing and disseminating news and information.

Photo Journalism:

It is a type of journalism in which photographs are used more than words. Someone who reports news using photographs is called photojournalist.

Print Making:

Print making is the process of making artworks by printing, normally on paper. Print making covers only the process of creating prints with an element of originality, rather than just being a photographic reproduction of a painting.

Radio:

Radio is an electronic audio medium for broadcasting programmes to the audience.

Rural Newspaper:

A rural newspaper has been defined as one that is primarily published from a rural area, devoting atleast 50 per cent of its contents to the problems relating to the rural community.

Situated Citizen Journalism:

Situated citizen journalism is an alternative and activist form of news gathering and reporting that functions outside mainstream media institutions, often as a repose to shortcoming in the professional journalistic field, that uses similar journalistic practices but is driven by different objectives and ideas and relies on alternative sources of legitimacy than traditional or mainstream journalism.

Soft News:

It is an event that takes place with notice. We can predict these events and they lack imperative and excitement.

CHAPTER 4

Diffusion and Adoption of Innovation

Diffusion of innovation seeks to explain how, why and at what rate new ideas and technology spread. It is one of the most important processes in cultural evolution.

Active Rejection:

It consists of considering adoption of the innovation (including even its trial), but then deciding not to adopt it.

Adopter Categories:

The members of social unit classified on the basis of the speed with which they adopt innovations of a certain type is called adopter categories. E.g., innovation based on agricultural research.

Adoption:

It is a mental process through which an individual passes from first hearing of an innovation to its final adoption.

Adoption Behaviour:

It is defined as the ability of the farmers to adopt a new technology.

Adoption Index:

Asking the farmers about how many improved practices recommended by the extension services they had adopted and for how many years is termed as adoption index.

Adoption Process:

The changes that take place within individuals with regard to an innovation from the moment that they first become aware of the innovation to the final decision to use it or not is termed as adoption process.

Adoption Research:

Research into the way that members of a social system adopt innovations is called as adoption research.

Adoption Quotient:

It is a ratio scale designed to quantify the adoption behaviour of an individual. It is developed by Chattopadhyay.

Authority Innovation Decision:

These are choices to adopt or reject an innovation that are made by relatively few individual in a system, who possess power, status or technical expertise.

Communication Channel:

It is the means by which knowledge spreads.

Compatibility:

It is the degree to which an innovation is perceived as consistent with the existing values, past experiences and need of potential adopters.

Complexity:

It is the degree to which the innovation is perceived as relatively difficult to understand and use.

Confirmation Stage:

Confirmation occurs when an individual or some other decision making unit seeks reinforcement of an innovation - decision already been made, or reverses a previous decision to adopt or reject the innovation if exposed to conflicting messages about the innovation.

Contingent Innovation Decision:

These are the choices to adopt or reject the innovation that can be made after a prior innovation decision.

Decision Making:

Decision making is a process comprising a sequence of stages with a distinct type of activity occurring during each stage. It includes all the considerations which play a role in making a choice from alternative action possibilities.

Decision Making Process:

The way in which decision making takes place is called decision making process. When describing a decision making process, different emphasis can be placed on the stages that one passes through, on the degree of rationality, on the relative participation of those involved in the decision making and on the rules that are followed.

Decision Stage:

The stage in which the decision occurs when an individual or some other decision making unit engages in activities that lead to a choice to adopt or reject the innovation.

Diffusion:

It is a process by which an innovation is communicated through certain channels over a certain period of time among the members of a social system.

Discontinuance:

Discontinuance is a decision to reject an innovation after having previously adopted it.

Disenchantment Discontinuance:

It is the decision to reject an idea as a result of dissatisfaction with its performance.

Dissonance:

It may be defined as the uncomfortable state of mind or a individual seeks to avoid a state of internal equilibrium.

Early Adopters:

They are localite and are a more integrated part of the community. Because, early adopters are not too far ahead, the average members of the community can comprehend their activities relating to adoption of the innovation. They are literate, have large size enterprise, high income, more participative and maintain good contact with cosmopolite sources of information.

Early Majority:

They adopt new ideas just before the average members of the community. They are neither very early nor relatively late to adopt an innovation. They are deliberate and take longer time to make the decision to adopt, in comparison to the innovators and early adopters.

Explicit Knowledge:

Explicit knowledge refers to all aspects of formal, systematic, recorded, communicated and shared knowledge that is made accessible through a variety of information delivery systems.

Gap Analysis:

It indicates the extent to which technologies have not been adopted.

Heterophily:

It is the degree to which pairs of individuals who interact are different in certain attribute.

Homophily:

It is the degree to which pairs of individuals, who interact are similar in certain attributes such as belief, education, social status and values.

Impersonal Localite:

Here the channels of communication are from outside the social system of the receiver and at the same time no personal face-face contact is involved.

Implementation Stage:

Implementation occurs when an individual or other decision making unit puts an innovation into use.

Innovation Decision Period:

It is the length of time required to pass through the innovation decision process.

Innovation Decision Process:

It is the mental process through which an individual passes from first knowledge of an innovation to forming an attitude towards the innovation to a decision to adopt implementation of new idea and confirmation of his decision.

Innovation:

Innovation is an idea or practice or object that is perceived as new by an individual or social system. It can also be described as the ideas, methods or objects that are considered new for the individual.

Innovativeness:

It is the degree to which an individual or a unit of adoption is relatively earlier in adopting new ideas than other members of social system.

Innovators:

They are venturesome and first to adopt a new idea, much ahead of other members in the community. They are generally very few in number. They may deviate from the social norm and may be viewed as deviants by others.

Knowledge Stage:

Knowledge occurs when an individual or other decision making unit is exposed to an innovation's existence and gains some understanding of how it functions.

Laggards:

They are traditional and the last to adopt an innovation. By the time the laggards finally adopt an innovation, it may already have been superseded by a more recent idea which the innovators are already using.

Late Majority:

They are cautious and skeptical, and adopt new ideas just after the average members of the community. They adopt mainly because people have already adopted the innovation and are getting the benefits out of it. They have low level of education, low level of participation and depend mostly on localite sources of information.

Observability:

It is the degree to which the results of innovations are visible to others.

Optional Innovation Decision:

This is a innovation decision where the adopting individual has the complete role or responsibility of the decision.

Over Adoption:

It is defined as the adoption of innovation by an individual when experts feel that he/she should rejects.

Passive Rejection:

It is also called as non-adoption which consists of never really considering the use of the innovation.

Personal Cosmopolite:

They are the extension agents of various organization and are important in changing the farmers from traditional to modern.

Personal Localite:

They are the local leaders and local people who belong to the own social system. Personal localite channels are important in traditional social system.

Persuasive:

Convincing, influencing attitudes and behaviour in a direction considered by the source is termed as persuasive.

Persuasion Stage:

Persuasion occurs when an individual or some other decision making unit forms a favourable attitude towards the innovation.

Predictability:

It refers to the degree of certainty of receiving expected benefits from the adoption of an innovation.

Preventive Innovation:

Preventive innovation is an idea that an individual adopts at one point in time in order to lower the probability that some future unwanted event will occur.

Rate of adoption:

It is the relative speed with which an innovation is adopted by members of a social system.

Rejection:

It is a decision not to adopt an innovation.

Relative Advantage:

It is the degree to which the innovation is perceived as being better than the idea it supercedes.

Replacement Discontinuance:

It is a decision to reject an idea in order to adopt a better idea that supersedes it.

Selective Exposure:

When the people expose themselves to messages selectively, such exposure is called selective exposure.

Social Marketing:

Social marketing is the application of commercial marketing strategies to the diffusion of non-profit products and services.

Selective Perception:

If an individual's perception of a certain event, issue, person or place could be influenced by one's latent beliefs, attitudes, wants, needs or other factors, then it is called selective perception.

Selective Retention:

All information is not retained by the individuals. People generally tend to retain that information in which they have some interest and which they consider to be important. Such thing is called as selective retention.

Social System:

It is defined as a set of interrelated units engaged in joint problem solving approach to accomplish common goal.

Tacit Knowledge:

It is highly personal, created by doing, trial, error, reflection and revision.

The Critical Mass:

The critical mass occurs at the point at which enough individuals have adopted an innovation so that the innovation's further rate of adoption becomes self-sustaining.

The Diffusion Effect:

Not only does change agent effort have a different effect at different points in the sequence of an innovation's rate of adoption, but the system's self generated pressure towards adoption also change as an increasing proportion of the members of the system adopt. This increasing pressure from interpersonal networks may be termed as the diffusion effect.

Trialability:

It is the degree to which the innovation may be experimented on a limited basis.

Uncertainty:

Uncertainty is the degree to which a number of alternatives are perceived with respect to the occurence of an event and the relative probability of these alternatives.

CHAPTER 5

Training and Training Methods

Training is teaching, or developing in oneself or others, any skills and knowledge that relate to specific useful competencies. Training has specific goals of improving one's capability, capacity, productivity and performance. Training and development can be described as "an educational process which involves the sharpening of skills, concepts, changing of attitude and gaining more knowledge to enhance the performance of employees". In Agricultural Extension, training of farmers and extension workers is atmost important one. Different methods are followed for training both the extension workers and farmers.

Brain Storming:

This is a group discussion technique in which the members generate as many ideas as possible on a specific topic without restriction including that of practicability.

Buzz Session:

Buzz session comprises of large group which is divided into smaller ones and a topic is discussed within a limited period of time. This technique involves dividing a large group into much smaller ones, in which a topic is discussed within a limited period of time. The groups are expected to produce ideas, opinions, questions etc.

Case Study:

In this method, a written case or a problem situation is presented to the participants in a programme for careful study and examination from all facets, so as to enable them to exercise their analytical, synthetical and decision making powers.

Clinic:

Clinic is a meeting of a group of people with common interests. It is established for the purpose of diagnosing, analyzing and seeking solutions to specific problems.

Colloquium:

This is a modified version of panel discussion, in which three or four resource persons discuss a specific topic. The audience are free to express their opinions, raise issues and ask questions. In colloquium, there is more of audience participation.

Colloquy:

A modified panel method in which there is a good degree of audience participation.

Committee:

A small group of individuals appointed or elected to perform a task that cannot be done efficiently by the entire group.

Conference:

Conference is a large event for a large gathering of hundreds or even thousands of people, to engage in a discussion with the aim of accomplishing a limited task within a limited period of time. It is an assembly of different organizations, departments for discussing issues of common interest.

Convention:

A convention is an assembly of representatives or delegates from local units of a parent organization who have a common interest or an assembly of representatives or delegates from different organizations or professions who have a common interest.

Counseling:

A form of mutual discussion, in which the expert's function is to help clarify the relationship of the client with him or herself and with others.

Curriculum Development:

Curriculum development consists of the construction, implementation and continuous updating of the teaching programmes offered by an educational institution.

Ex Post Training Evaluation:

This type of evaluation is undertaken after completion of training project, usually after full project benefits, and impact are expected to have been realised.

External Training:

Since extension organizations are not able to meet all the training needs of extension personnel within the organizations, they depute their personnel to other training institutions who have specialized in different subject matter areas.

Focus Group Interview:

A group of people who possess certain characteristics provide data of a qualitative nature in a focussed discussion is termed as focus group interview.

Formative Training Evaluation (FTE):

FTE pertains to the implementation stage of the training programme and helps in day-to-day decision making.

Foundation Training:

This type of training is usually provided at an early stage of service life. Every staff member needs some professional knowledge about various rules, regulations, and procedures of the organization, financial transactions, administrative capability, communication skills, learning ability, coordination and cooperation among institutions and their linkage mechanism, report writing and so on. Thus training is made available to the employees to strengthen the foundation of their service career.

Group Interview:

This method is used to obtain community level information. It provides access to a larger body of knowledge and provides an immediate cross-check on information, as it is received from several people in the group.

Individual Interview:

Individual interview is the one where the interviews are conducted with purposively selected individual respondents.

Induction Training:

It is the training given to new extension personnel after they have been employed and before they are assigned to work in particular area.

Inservice Training:

It includes all forms of training for professional extension personnel during their period of employment by the extension service.

Institute:

An institute is a series of meetings arranged for a group of persons who come together to receive instructions or information about, a specific field of work.

Job Analysis:

It is a process of gathering, analyzing and recording the relevant information about the functions and operations involved in the performance of a job.

Key Informant Interview:

Key informants are the people who are considered experts in a given situation because of their professional knowledge or their position of influence in the community or organization.

Management Game:

It is the training method that deals with certain specific aspects of business or administration. It is a simulated exercise representing, as closely as possible the constraints and pressures of the day-to-day work environment of the participants.

Microlab:

A microlab is organized at the beginning of any programme to provide reflection of all aspects likely to be covered in the programme. It is full of physical activity with the major purpose of promoting communication among the group members and for helping them overcome barriers due to fear, status consciousness, etc.

Ongoing Training Evaluation:

It is the training evaluation which is undertaken during the course of a training activity.

Orientation Training:

It is usually given to the newly appointed extension personnel. It provides an introduction to public employment, and attempts to provide answers to questions which a newly recruited person is likely to ask.

Overseas Training:

It is used to enable extension personnel to acquire expertise in areas in which such expertise is not available indigenously to the required degree.

Panel Discussion:

A panel or a group of three or four experts in a specific area of specialization may be invited to address a group of trainees on a particular subject is called panel discussion.

Panel:

This is the moderated meeting in which a limited number of experts give short presentations on some subject. A moderator oversees the question-answer period and sums up the meeting.

Pre-Service Training:

It refers to the professional training prior to appointment to the extension service.

Refresher Training:

This type of training is mainly for refreshing the knowledge and skills of trainers for imparting to trainees. This training is usually imparted in the later career of extension personnel.

Re-Training:

Re-training is an instructional effort designed to prepare an individual for a new assignment or new duties.

Role Playing:

Role playing is a simulation in which participants enact different roles to obtain insight into a behaviour, problem or situation. The participants can be made to act and re-live a particular situation to get a real feel of the roles that are actually called upon to play.

Round Table:

It is so named because experts sit around a round table to discuss a subject.

Seminar:

It is one of the most important forms of group discussion and is more formal in nature. The discussion leader introduces the topic to be discussed. The audience will discuss the subject to which ready answers are not available. A seminar may have two or more plenary sessions that enables to study in depth in specific areas under the guidance of experts. Seminar is a brief, opening presentation, often on provocative issues, and guide general discussion.

Semi-Structured Interview:

It is a form of guided interviewing where only some questions are predetermined. Many questions are formulated during the interview, as in the journalistic interview.

Sensitivity Training:

Sensitivity training is the training where the members are brought together in a free and open environment in which participants discuss themselves and interact. The discussion is lightly directed by a behaviour expert, who creates the opportunity for participants to express their ideas, beliefs and attitudes.

Simulation Games:

This is a learning exercise in the form of a game which contains elements of real life situation. This gives the participants a practice in dealing with life situations.

Simulation:

Simulation is a technique for understanding processes by imitating a situation which happens in reality or can happen.

Skill:

Skill is the ability to do things, to effectively apply knowledge and personal aptitudes and attitudes in work situations.

Small Group Training:

It is a technique of imparting specific skills to a group of people who need them by creating an appropriate learning situation.

Summative Training Evaluation (STE):

STE relates to final evaluation of the objectives of the training programme.

Symposium:

It is a meeting in which a small number of resource persons present short prepared papers on a given topic. It is used mainly for gathering information at professional level where 2-5 resource persons give short prepared papers on a given topic with no audience participation.

Syndicate Method:

A syndicate is primarily a study group, the members of which represent the principal sub-groups participating in the programme. A problem is assigned to the syndicate for study and solution.

Task Force:

It is a small team of experts complete in its requirement of specialties for examination of a subject.

T-Group:

It is a method of changing behaviour through instructed group interaction. It provides insights to one's own behaviour, sensitivity to others and improvement in behavioural skills but has not yet found any significant application in extension training in India.

Training:

Training is a process by which the desire, knowledge, attitude, skill and ideas are inculcated, fostered and reinforced in an organism. It is the process of acquiring specific skills to perform a better job.

Training by Overseas Experts:

Overseas experts are at times invited to train extension personnel.

Training Evaluation:

Training evaluation is a systematic process of collecting information for and about a training activity which can then be used for guiding decision making and for assessing the relevance and effectiveness of various training components.

Training for Professional Qualification:

This type of training is utilized to enable extension personnel to acquire graduate and post graduate qualifications in State Agricultural Universities and institutes deemed as universities.

Training Review:

Training review is defined as the systematic, periodic effort to integrate the results of all the training programmes and activities undertaken by an extension organization including deputation of its personnel to other training institutions and management of training institutions falling within its purview, with a view to make training more productive and useful to extension personnel.

Training Strategy:

Training strategy is not a static blueprint but is open to clarification, modification and further development through experience with it.

Working Party:

Working party is an informal committee of experts engaged in preliminary work pertaining to a problem.

Workshop:

A workshop is a cooperative gathering of individuals who discuss, learn and apply practical skills under expert supervision. It is a long meeting from one day to several weeks, involving all the delegates in which the problems being discussed are considered by delegates in small private groups. Workshop is gathering for providing practical experiences.

CHAPTER 6

Participatory Methods and Programme Planning in Extension

Programme is an overall, long ranged schedule or broad outline of things that need to be done and methods of doing them. Programme planning is a continuous series of activities leading to the development of a definite plan of action to accomplish particular objectives. Extension programmes are implemented at the local level. Local people should, therefore, be involved all through, from programme formulation to programme implementation.

Activity Slack (AS):

It is the slack associated with an activity.

Activity:

Activity is a definite part of the total undertaking. It must be accomplished in order to complete the projects.

Advisory Committee:

An advisory committee may be defined as a group of people organised from among the community to serve as an advisory group to the extension workers, who are responsible for the development and execution of an extension education programme at the block level.

Agricultural Technology Information Centre (ATIC):

ATIC is a "single window" support system linking the various units of a research institution with intermediary users and end users (farmers) in decision making and problem solving exercise. It is a component of NATP to be implemented by Agricultural Universities and ICAR institutes.

Aim:

Aim is generalized and broad statement of directions with respect to given activities.

Backward Pass:

Backward pass is a procedure in which Latest Starting Point (LST) values are computed. It starts with the end event and ends with the first event of the project.

Calendar of Work:

A calendar of work is a plan of activities to be undertaken in a particular time sequence.

Critical Activities:

These are activities which fall on the critical path.

Critical Events:

These are events which fall on the critical path.

Critical Path Method (CPM):

CPM is a technique of project management useful in the basic management function of planning, scheduling and controlling. CPM is a deterministic model and activity oriented network which considers the costs for performing various activities of a project.

Critical Path:

It is the longest path for completing the project. All critical activities and critical events lie on this path.

Direct Matrix Ranking:

This method enables the participatory team to prioritize from a list of criteria for a certain object.

Direct Observation:

Direct observation is the observation where the objects, events, places, processes, relationships or people are systematically observed and recorded those observations promptly.

Dummy Activity:

It indicates the very logical relationship between preceding activity and succeeding activity relationships. It does not require any resource - money, men, material and time.

Earliest Starting Time (EST):

It is the time at which the activity commences. This is also the time before which the activity cannot be started.

Efficiency - cum - Performance Audit:

It is the examination of the systems, procedures, planning, implementation and operational performance of programmes, activities, etc. with a view to bringing out, among other things, weaknesses and deficiencies as also lapses of various types including those relating to individual transactions, for appropriate actions.

Evaluation:

Evaluation is the process of determining the extent to which one has been able to attain the objectives. Evaluation can also be defined as assessing the worth of certain things, which may be objects, persons, processes, achievements and the like.

Event Slack:

The slack time associated with an event is called as event slack.

Event:

It is a definite state in the progress of the project. Event indicates the beginning and end of an activity.

Extension Programme:

Extension Programme is a plan of the goals and target groups an extension organization or extension unit tries to reach, and of the ways it will try to reach them. It is a statement of situation, objectives, problems and solutions. It forms the basis for extension plans.

External Validity:

It is the extent to which an evaluation is generalizable to other similar events which persist a problem in education evaluation.

Farmer Participatory Research (FPR):

FPR is an approach, which involves encouraging farmers to engage in experiments in their own fields so that they can learn, adopt new technologies and spread them to other farmers.

Felt Need:

It is the consciously recognized need i.e the need of which the people are aware.

Formative Evaluation:

The formative evaluation tries to increase the effectiveness of an extension campaign by collecting information on reactions of the target group and experiences of the extension agents during implementation of the programme.

Forward Pass:

Forward pass is the procedure in which Earliest Starting Time (EST) values are computed. It starts with the first event and ends with the last event of the project.

Generalization:

The formulation of general concepts based on a number of specific observations is called generalization.

Goal:

Goal is the distance in any given direction one expects to go during a given period of time. It is the end towards which actions are moving. E.g., extension programmes.

Innovation Assessment:

Innovation assessment helps to assess and prioritize possible options for development activities according to some criteria which may be the amount of benefit for the community, degree of community participation, sustainability of the project, equitability of distribution of benefits, cost of implementation, gestation period, technical feasibility etc.

Internal Validity:

It is the correspondence between reliability and evaluation, unless the evaluation procedures results which relate to processes.

Latest Starting Time (LST):

It is the time beyond which the starting of an activity would not be delayed without introducing delay in the project completion.

Matrix Ranking/ Preference Ranking:

It is the method of ranking and scoring which reveals priorities and preferences of the villagers. It provides opportunities to all the village people to value and rank the problems and prioritize themselves.

Mission:

A powerful group of front-rank experts charged with attainment of specific objectives and goals in a 'thrust' area at the national level is called mission.

Need:

Need implies a gap between what is the existing situation and what ought to be the desirable situation.

Network:

Network may be defined as the diagrammatic representation of various activities of the project, showing the preceding and immediate succeeding activity relationships, their time and cost requirements in accomplishing a particular activity.

Objectives:

An objective is an expression of the end or direction of movement towards which efforts are directed.

Objectivity:

It refers to the degree to which the process of measurement is free from biasness.

Ongoing or Process Evaluation:

Ongoing or process evaluation means evaluation when the work is in progress. It enables the evaluator and the stakeholders to develop a better understanding of the functioning of the programme.

Outcome Evaluation:

Outcome evaluation means evaluation after completion of the work. It enables the participants to apply the understanding which they develop in the process evaluation to assess which of their goals are being achieved, and how well this is being done.

Pairwise Ranking:

In this method, the socially relevant items relating to problems, preferences etc. are separately presented to the respondents in pairs in all possible combinations.

Participatory Appraisal of Natural Resources (PANR):

PANR is concerned with local people's appraisal who live in proximity with such resources and have a close relationship with them.

Participatory Evaluation:

Participatory evaluation is designed to help people to help themselves and improve their programmes using a form of self-evaluation and reflection.

Participatory Mapping and Modeling:

This involves constructing, on the ground or on paper, maps, or models, using locally available materials such as sticks, stones, sand, soils, grasses, wood, rangoli powders, coloured chalk, pen and paper etc. The person who holds the stick tasks about what is most important to him or her.

Participatory Methods:

Participatory methods are discovery learning process, which do not define the problems and solutions for the people, but make things visible so that the people could define their own problems and develop their own solutions.

Participatory Rural Appraisal (PRA):

PRA is the process of involving local people in the analysis and the interpretation of a rural situation.

Participatory Technology Development:

The process of combining local famers knowledge and skills with these of external agents to develop site specific and socio-economically adopted farming technologies is called PTD.

People's Participation:

It is defined as the process of giving priority to local people's perspective in identifying and analyzing their problems and opportunities and improving the situation through their self mobilization.

Performance Appraisal:

It is a system of periodical appraisal of employees to assess their performance in terms of output quality and interpersonal behaviour.

Performance Budget:

Performance budget is one which presents the purposes and objectives for which funds are required, the costs of the programmes proposed for achieving these objectives, and quantitative data measuring the objectives and accomplishments, and work performed under each programme.

Plan of Work:

A plan of work is an outline of activities arranged so as to enable efficient execution of the programme. It is a statement of activities to be undertaken by an individual, a group of people or an organization, within a given period of time, to carry out the recommendations in the programme.

Plan:

A plan is a predetermined course of action. Plan is usually an annual schedule of work. It answers the questions of what, why, when, where and by whom and how the work is to be done.

Planning:

Planning is a process which involves studying the past and present in order to forecast the future.

Policy:

Policy is a form of directed action which indicates as clearly as possible what one wants to achieve, how one wants to do it and how much time will be taken to achieve the set goals.

Practicability:

It refers to the degree to which the evaluation procedures are convenient to use, understand and be appropriate for application by the user.

Preference Ranking:

This allows the participatory team to determine quickly the main problems or preferences of individuals and enables the priorities of different persons to be easily compared.

Principle:

A principle may be defined as a statement of policy to guide decisions and actions in a constant manner.

Problem:

Problem is a condition that the people after study, with or without help, have decided needs changing. It is the situation in which a person who has a goal does not know how to reach that goal.

Process:

Process refers to a course of procedures, something that occurs in a series of actions or events conducting to the desired end.

Programme Evaluation and Review Technique (PERT):

PERT has been used for research types of programmes, especially when there are a number of uncertainties in the development of new technologies. This technique is originally developed in 1958 to facilitate the planning, scheduling of the Polaris Fleet Ballistic Missile Project of the U.S. Government.

Programme Evaluation:

Programme evaluation is the determination of the extent to which the desired objectives have been attained or the amount of movement that has been made in the desired direction.

Programme Planning:

It is a continuous series of activities leading to the development of a definite plan of action to accomplish particular objectives. Programme planning is a decision making process involving critical analysis of the existing situation and the problems, evaluation of the various alternatives to solve these problems and the selection of the relevant ones giving necessary priorities based upon local needs and resources by the cooperative efforts of the people both official and non-official with a view to facilitate the individual and community growth and development.

Programme Projection:

Programme projection means improved programme planning by looking more broadly, looking more deeply, looking longer or farther ahead and, necessarily, involving more people.

Programme:

A programme is the total educational job being done in a particular situation. It is an overall long ranged schedule or broad outline of things that need to be done and methods of doing them.

Programmed Instruction:

An instructional method in which the subject matter is divided into small units, some of which require an answer from the learner.

Project Formulation:

Project formulation is the process of studying a selected project further or developing a new project with reference to investment decisions.

Project Management:

Project management is defined as the planning, scheduling, and controlling of project activities to accomplish the desired objectives. It is the process of managing and directing the appropriate use of resources such as people, materials, money and time to complete a particular project according to the objectives set forth.

Project Selection:

Project selection is a careful study of each project idea in detail by Strength, Weakness, Opportunity and Threat (SWOT) analysis and choosing one of them for further consideration and development.

Project:

A project is a specification of work to be done or procedure to be followed in order to accomplish a particular objective. It is a single item of the annual plan containing the method and solution of single selected problem.

Ranking:

Ranking means placing some items in order. These are useful for collecting sensitive information like income and wealth.

Rapid Rural Appraisal:

It is an activity conducted on site by a multidisciplinary team which uses simple non-standard methods and local peoples knowledge to elicit, analysis and evaluate information and hypotheses about rural life and resources, that are relevant for taking action.

Reliability:

It refers to consistency of measurement, the problem or evaluation procedures and their results unless the details of evaluation and procedures of programme being evaluated are recorded.

Report:

A report is a formal record of activities or performance. It generally reflects the achievement or otherwise of a programme.

Resource Map:

It is a diagram which shows mainly available natural resources such as topography, forest areas, degraded land, pastures, water resources, agricultural lands, which are the major elements for overall development plan.

Rural Development:

Rural development is a strategy to enable a specific group of people, poor rural women and men, to gain for themselves and their children more of what they want and need. It aims at improving the quality of life of the poor through technological intervention, forming, modifying or strengthening new and old institutions and organizations for building their capacities, providing better communication facilities and implementing government policies and programmes.

Seasonal Calendar and Activity Profile:

Seasonal constraints and opportunities which can be diagrammed month by month throughout the year is called seasonal calendar.

Seasonality Analysis:

The analysis which is applied to get an insight into seasonal variations in rural areas related to rainfall pattern, crop rotation, fodder and fuel availability, labour requirements, insect and pest attack, etc. is called seasonality analysis.

Simplicity:

It refers to the extent to which the entire processes of the evaluation facilitate the user for which the purpose of evaluation intended to.

Situation:

It is a statement of affairs that includes the cultural, social, economic and physical conditions in which a particular group of people find themselves at a given period of time.

Slack:

It is the more time (spare time) available in commencing with an activity.

Social Map:

It is a diagram depicting village layout, roads, human habitation, schools, post office, temple, anganwadi centre, cattle, ownership of assets etc.

Solution:

Solution is a course of proposed action to change in unsatisfactory condition to one that is more satisfying.

Strategy:

A way to achieve clearly specified goals with a combination of means and in a certain time period is called as strategy.

Summative Evaluation:

Summative evaluation tries to measure the changes achieved with a completed extension programme in order to use this experience to improve other extension programmes.

Sustainability Analysis:

Sustainability analysis is highly useful for the periodic review of development activities. Without additional collection of information, this matrix sharpens the group's analytical skills and brings out important issues.

Time Line:

It is a method of visualizing key historical events and major changes perceived by the villagers. Eg., Flood, soil erosion, climate changes, culture, etc. The best informants are the old age people of the village.

Historical Analysis:

Historical analysis includes detailed accounts of the past, of how things have changed, particularly focusing on relationships and trends.

Target Group:

A general indication of the more or less related group(s) which the extension agent or the extension organization wants to reach or change is the target group. The target group can refer to individuals or organizations.

Transact Walk:

It is an exploratory walk which is undertaken by a multidisciplinary team of scientists along with village people to observe, cross check and critically examine and record minute details of a particular area physically.

Transect Diagram:

Transect diagrams may be drawn by making the team walk through the village and surrounding areas, observing, asking, listening and discussing problems and opportunities while walking.

Unfelt Need:

It is the unrecognized need i.e. the need of which the people are unaware.

Validity:

Validity refers to the extent to which the results of the evaluation procedure serve the particular validity for which that are intended.

Venn and Network Diagram:

Venn diagram is the PRA tool that is applied to represent the role of individuals/institutions and the degree of their importance in decision making and their performance. It involves the use of circles of paper or card to represent people, groups and institutions. These are arranged to represent degree of linkages and interaction between individuals and institutions.

Overlap indicates the flow of information and distance on the diagram represents lack of contact.

Wealth Ranking:

Wealth ranking is a method in which village people jointly determine the relative wealthiness taking into account all asserts, sources of income, and liabilities of an individual family. Wealth ranking is used to classify households according to relative wealth or well being.

CHAPTER 7

Extension Education, Research Design, Extension Pluralism and Group Led Extension

Extension is education and its purpose is to bring about desirable changes in the knowledge, skills and attitudes of the people. Extension means to extend, to spread or to disseminate useful information and ideas to rural people outside the regularly organized schools and class rooms for bringing out the social and cultural development.

Action Research:

Action research is a form of social research which aims at better insight into the problem by learning from the experience gained in an attempt to solve this problem.

Actor:

A person who plays an active role in the Agricultural Knowledge and Information System (AKIS).

Adult Education:

Adult education is the process whereby persons who did not attend school on a regular and full time basis, undertake sequential and organized activities with conscious intention of bringing about changes in information knowledge, understanding or skills appreciation and attitude for the purpose of identifying and solving personal and community problems.

Agricultural Development:

Change in agricultural production techniques and in farming systems towards a more desirable situation, usually one in which farmers use more agricultural research findings and in which there is less subsistence and more market - oriented agriculture.

Agricultural Knowledge and Information System (AKIS):

The persons, networks and institutions and the interfaces between them, which engage in or manage the generation, diffusion and utilization of knowledge and information, which potentially work synergistically to improve the goodness of fit between knowledge and environment, and the technology used in agriculture.

Agricultural Technology Management Agency (ATMA):

ATMA is one of the innovations-technology dissemination components of National Agricultural Technology Project (NATP). It functions as a registered society at District Level and serves as a focal point for integrating research and extension activities and helps in decentralizing decision making to the district level.

Assumptions and Ideas:

Assumptions and ideas consist of basically untestable premises about the nature of selected aspects of social life.

Broad Based Extension System (BBES):

BBES is also an approach of extension for increasing the income of farmers. The idea of this system is bringing a number of farm activities in addition to the crop production under the umbrella of agricultural extension.

Cattle Development Project:

The cattle development project was launched in 1964 which is envisaged the provision of a package of improved practices to the cattle owners to effect a break through in milk production.

Community Development:

Community development is a process by which the efforts of the people themselves are united with those of the governmental authorities to improve the economic, social and cultural conditions of the nation and to enable them to contribute fully to national progress.

Community Development Programme (CDP):

This programme was launched on 2nd October 1952. The central objective of the CDP was to secure the total development of the material and human resources of rural areas and to develop local leadership and self-governing institutions.

Concepts:

Concepts are simply abstractions that serve to organize our thoughts and experience into manageable mental entities. Concepts vary greatly in terms of their precision, scope, clarity and acceptability.

Continuing Education:

It refers to updating and enhancing the capabilities of a person in order to increase his effectiveness and efficiency in a particular job. It usually starts sometime after the formal education and continues even while the person is generally employed.

Contract Farming:

Contract farming refers to contractual arrangement between farmers and companies whether oral or written, specifying one or more conditions of production and/or marketing of an agricultural product.

Control of Extraneous Variable:

It means the influence of independent variable extraneous to the purpose of the study that are minimized/nullified or isolated.

Cyber Extension / e-Extension:

Cyber extension is an agricultural information exchange mechanism over cyber space, the imaginary space behind the interconnected computer networks through telecommunication means. It utilizes the power of networks, computer communications and interactive multimedia to facilitate information sharing mechanism.

Decentralization:

Decentralization means distribution of functions and powers from a central authority to regional and local authorities.

Democratic Decentralization:

Democratic decentralization means that the government which has derived its authority from the people redistributes it to some extent to the people for decision and action at the local level.

Development:

Development means growth or maturation. It is the positive changes. It implies gradual and sequential phases of change.

Development of Women and Children in Rural Areas (DWCRA):

DWCRA is launched on 1982-83 as pilot project in 50 districts chosen on the criteria of high infant mortality rate and low female literacy. The main objective of DWCRA is to provide rural women with productive income-generating assets and credit and enhance their skills.

Education:

Education is the process of developing capabilities of the individuals so that they can adequately respond to their situations. It is the process of bringing desirable changes in the behaviour of human beings.

Error Variance:

Error variance is the variability of measures generated by random fluctuations whose basic characteristic is that they are self compensating, varying now this way, now that way, now positive, now negative, now up and now down tending to balance each other.

Extension:

Extension is a continuous process designed to make the rural people aware of their problems and indicating to them the ways and means by which they can solve them. It thus involves not only education of the rural people in determining their problems and methods of solving them, but also inspiring them towards positive action in doing so.

Extension Education:

Extension is the type of education which is stretched out to people in the rural areas far and near, beyond the limits of the educational institutions to which the formal type of education is usually confined. Extension education is an applied behavioural science, the knowledge of which is applied to bring about desirable changes in the behavioural complex of human beings usually through various strategies and programmes of change and by applying the latest scientific and technological innovations.

Farmer Led Extension:

Farmer led extension in agriculture is the process through which individuals or groups within a given locality discover or develop and apply improved ways of managing the available resources, building on and expanding the boundaries of their indigenous knowledge.

Farmer Field School:

The Farmer Field School is a form of adult education, which evolved from the concept that farmers learn optimally from field observation and experimentation.

Farming Systems Research (FSR):

This is a type of research through which an inter disciplinary research team tries to gain as complete knowledge as possible of the existing farming system in order to assess whether a new technology helps farmers in achieving their goals under those environmental conditions.

First Line Extension System:

The first line extension system comprises of the Indian Council of Agricultural Research (ICAR) institutes and the State Agricultural Universities.

Formal Education:

Formal education is the highly institutionalized, chronologically graded and hierarchically structured 'education system', spanning lower primary school and the upper reaches of the university.

Front Line Demonstration:

The aim of front line demonstration is demonstrating the production potentialities of newly released and pre-released production technologies of cereals, pulses and oilseeds on farmers fields. These are called frontline demonstrations because the technologies are demonstrated for the first time being fed into the main extension system.

Gram Panchayat:

It is the lowest level tire of panchayat. It is at the village level.

Gram Sabha:

Gram Sabha means a body consisting of persons registered in the electoral rolls relating to a village comprised within the area of gram panchayat.

High Yielding Varieties Programme (HYVP):

The HYVP was launched in Kharif 1966-67. The sole objective was to increase the total food production by utilizing high yielding seeds of selected crops.

Index:

Index is the number on the scale given to the variable involved.

Indigenous Knowledge:

The knowledge held in the rural society, usually based on the experience of many generations and unique to each cultural group.

Informal Education:

Informal education is the lifelong process by which every person acquires and accumulates knowledge, skill, attitude and insights from daily experiences and exposure to the environment at home, at work, at play etc.

Intensive Agricultural District Programme (IADP):

It was launched during Kharif 1960. The main goal was to achieve rapid increases in agricultural production through a concentration of financial, technical, extension and administrative resources.

Integrated Child Development Services Programme (ICDS):

This scheme was launched on 1975-76 in 33 community development blocks on an experimental basis. This particular programme is mainly designed to facilitate and promote "total development" of the child by making available of integrated package of services comprising mutually reinforcing components of health, nutrition and preschool education.

Integrated Rural Development Programme (IRDP):

The programme was started during 1978-79. The main aim of this programme is providing income generating assets and employment opportunities to the rural poor to enable them to rise above the poverty line once and forever.

Integrated Tribal Development Agency (ITDA):

This was a special programme for the development of selected tribal areas which was initiated during 1970-71. This programme is also called as Tribal Areas Development Programme (TADP). The main objective of the programme is socio-economic development of tribal communities through income generating schemes allied with infrastructure development programmes and protection of the tribal communities against exploitation.

Institution Village Linkage Programme (IVLP):

Technology Assessment and Refinement (TAR) through Institution Village Linkage Programme (IVLP) is one of the programmes of the ICAR emphasizing on technology transfer. IVLP is a programme for generating appropriate technologies by refining and assessing innovation generated by scientists about their respective disciplines.

Joint Liability Group (JLG):

JLG is an informal group comprising preferably of 5 to 10 individuals coming together for the purposes of availing bank l oan either singly or through the group mechanism against mutual guarantee.

Knowledge Management:

Knowledge management can be defined as the fact or condition of knowing something with a considerable degree of familiarity acquired through experience, association or contact.

Krishi Vigyan Kendra:

Krishi Vigyan Kendra is also called as Agricultural Science Centre. It is an innovative institutions for imparting vocational training to the practicing farmers, school dropouts and field level extension functionaries.

Learning:

Learning is the process by which an individual, through his own activity attains a change in his behaviour.

Mahil Samridhi Yojana (MSY):

It was started on 2nd October, 1993 by National Minorities Development and Finance Corporation, Ministry of Minority Affairs, Govt. of India (NMDFC). The main objective of implementing this programme is to build the capacity of minority women through increasing income generating skills/activities; encouraging them to form SHGs and provide them micro credit support; setting up of production based activity suitable to minority women.

Manipulation:

Manipulation is the systematic attempt of individuals and groups to influence the behaviour, attitudes and opinion of other people without being aware of it.

Market Led Extension:

In market led extension, the extension service is provided based on market demand.

Market Research:

Market research is the research which is directed at gaining insight into markets for products.

Monthly Workshop:

The monthly workshop or monthly zonal workshop is held at monthly interval. The main objective is to train Subject Matter Specialist (SMS), particularly in the production recommendations that extension workers will teach to farmers over a four-week period that usually starts some two or three weeks after the workshop.

Fortnightly Training:

It is held at fortnightly interval. It is the chief means of continuously upgrading and updating the professional skills of extension workers and of infusing them with confidence to meet farmers.

National Agricultural Innovation Project (NAIP):

The project was started during July 2006 to facilitate accelerated and sustainable transformation of agriculture in support of poverty alleviation and income generation through technological innovations by the public research organizations in partnership with farmer's group, the private sector, the civil society organizations and other stakeholders.

National Agricultural Technology Programme (NATP):

This programme was launched in November 1998 as a pilot project. The overall objective of NATP was to revitalize the agricultural technology generation, assessment, refinement and dissemination systems.

Non-Formal Education:

Non-formal education is an organized, systematic, educational activity carried on outside the framework of the formal system to provide selected types of learning to particular sub-groups in the population, adults as well as children according to their need.

On - Farm Research (OFR):

On farm research is an adaptive research which is conducted on the farmers fields by the farmers with the support from Scientists/ Subject Matter Specialists. It is conducted with a farming systems perspective. The main objective of OFR is to identify existing inputs or practices that might help to solve major problems of many farmers in a defined study area.

Open Learning / Distance Education:

It is the methodology in which the learner does not go to educational institution but the education comes to his doorstep.

Panchayat Samiti:

It is the middle level tire of panchayat. It is at the block level.

Philosophy:

Philosophy is the pursuit of wisdom, a body of general principles or laws of a field of knowledge.

Plan:

Plan is the overall scheme or programme of research. It includes an outline of everything that an investigator will do from writing the hypothesis and their operational implication to the final analysis of data.

Prime Minister's Rozgar Yojana (PMRY):

The PMRY was launched on 2nd October, 1993. The main objective of the scheme was to provide easy subsidized financial assistance to educated unemployed youth for starting their own enterprises in manufacturing, business and service and trade sectors.

Principle:

Principles are generalized guidelines which form the basis for decision and action in a consistent way.

Private Extension:

The process of funding and delivering the extension services by private individual or organization is called private extension..

Privatization:

It is the act of reducing the role of government or increasing the role of private sector in an activity or in the ownership of assets.

Public Private Partnership (PPP):

PPP is a mode of implementation of government programme / schemes or developed technologies by government agencies in partnership with private sector.

Randomization:

It is the method of selection in which all the individuals will get equal chance of being selected.

Rashtriya Mahila Kosh (RMK):

It was set up in March 1993. The main objective of this programme is to promote or undertake activities for the promotion of or to provide credit as an instrument of socio-economic change a nd development through the provision of a package of financial and social development services for the development of women.

Recurrent Education:

Recurrent education is a comprehensive and flexible compulsory educational system which combines the present formal educational branches and the various types of adult education.

Respondents:

The people who have answered the questions which is asked by an interviewer for a social survey are called as respondents.

Rural Development:

It is a process which aims at improving the well being and self realization of people living outside the urbanized areas through collective process.

Swarnajayanthi Gram Swarozgar Yojana (SGSY):

The programme was started from 01.04.1999. The basic objective of the SGSY is to bring the assisted poor families (Swarozgaris) above the poverty line by providing them income generating assets through a mix of bank credit and governmental subsidy.

Self Help Group (SHG):

Self help group is a small, economically homogeneous affinity group of rural poor, voluntarily formed to save and mutually agree to contribute to a common fund to be lend to its members as per group decision for socio-economic development.

Strategic Research and Extension Plan (SREP):

The SREP is the basic document which includes the development activities that need to be carried out, in which manner and whom it has to be done.

Structure:

Structure is the outline of a scheme.

Subject Matter Specialist (SMS):

An extension agent who has studied one scientific discipline or one specific problem area are regarded as subject matter specialists.

Teaching:

Teaching is the process of arranging situations in which the important things to be learned are called to the attention of the learners, their interest developed, desire aroused and action promoted.

Technology:

Technology is the application of knowledge for practical purposes.

Theory:

Theory is a set of proportional statements which present the distinct picture of the phenomena by determining relations among variables with the purpose of explaining and predicting the phenomena.

Variable:

A general indication in statistical research which occurs in a number of individuals, objects, groups, etc., and that can take on various values is called variable. E.g., age of an individual. Variables are relational units of analysis that can assume any number of designated sets of values.

Zilla Parishad:

It is the highest level or third tire of panchayat. It is at the district level.

4-H Clubs:

It is a global network of youth organizations whose mission is "engaging youth to reach their fullest potential while advancing the field of youth development. The 4-H Stands for Head, Heart, Hands and Health.

CHAPTER 8

e-Extension

e- Extension is the electronic delivery of extension service of the various institutions that provide a more efficient alternative to a traditional extension system for agriculture, animal husbandry, fisheries and natural resources sector.

Access Grid Videoconferencing:

The Access Grid is a dedicated videoconferencing room that is capable of very high quality display of multiple video sources and offers high speed connectivity.

Animation:

Animation is a technique of moviemaking, showing slightly different drawings in rapid succession, creating the illusion of continuous motion.

Blog:

Website with a regularly updated list of commentary and links to information on the Internet is termed as a blog.

Bookmark:

Bookmark is an address of a website that is stored in a computer's memory so that it can be revisited easily.

Byte:

A basic unit of electronic data storage, equal to 8 bits is byte. Files and computer memory are measured in bytes, kilobytes, megabytes or gigabytes.

Classroom Capture:

Class capture sometimes referred to as content capture, lecture capture or course capture. It is a system for recording (capturing) classroom based presentations and sharing them on the Internet.

Compact Disc - Read Only Memory (CD-ROM):

It refers to a storage medium that, together with a micro computer, offers, rapid access to a very large volume of data, including text, sound, computer graphics, animation slides and a limited amount of motion video.

Computer:

Computer is an electronic equipment designed to automatically accept and store input data, manipulate them and produce output results under the direction of a detailed step by step stored programme of instructions.

Computer Aided Design (CAD):

It is the use of computers for automatic design operations.

Computer Aided Instruction (CAI):

A general term that refers to teaching with the help of a computer. It is also called as Computer Aided Education (CAE).

Computer Aided Manufacturing (CAM):

It refers to the use of computers to automate manufacturing operations.

Computer Graphics:

This is the area of computer science which is concerned with the generation, manipulation and display of pictures with the aid of a computer.

Computer Network:

Computer network is a distributed data processing system in which multiple computers are linked together for the purpose of data communication and resource sharing.

Cyberspace:

It is a term for the shared imaginary reality of computer networks. It is also used as a synonym for the internet.

Database:

Database is an electronic information storage system which may contain social, technical or bibliographic information which can be searched and retrieved rapidly using appropriate software and computer.

Decision Support System:

Decision support systems are computer programmes that are different from conventional computer programmes as they solve problems by mimicking human reasoning processes, relying on logic, belief, rules of thumb, opinion and experience. It is the simulation models used to calculate feed rations, pesticide application and other on-farm calculation and advisory activities.

Disc:

Disc is a flat, circular plate coated with a magnetic material on which data can be stored by the process of magnetization.

Diskette:

Diskette is a low cost, thin, flexible magnetic disk storage device used on small computer systems. It is also called as a floppy disk.

e-commerce:

Commercial transactions between two parties on the internet is called e-commerce.

e-Education:

e-Education is a flexible form of education that allows participants with different schedules in different locations to interact and learn in a collaborative, online environment.

e-Governance:

e-Governance is the use of information communication technology (ICT) by the government, civil society and political institutions to engage citizens through dialogue and feedback to promote their greater participation in the process of governance of these institutions.

e-Learning:

e- Learning is learning which takes place as a result of experiences and interaction in an internet environment. It refers to the use of internet technologies to deliver a broad array of solutions that enhance knowledge and performance.

Electronic Book:

Electronic book is a digital book having the appearance of a traditional paper book but with its content stored digitally. e-books can be updated from a bookstore or a website that sells them.

Electronic Bookmark:

A function on an Internet browser that allows a user to return to a web page easily is electronic bookmark.

Electronic Business (e-Business):

It is the practice of conducting business through the Internet.

Electronic Magazine (e-Zine):

A magazine that is accessed online is electronic magazine.

Electronic Mail (e-Mail):

A mail sent from one computer to another is electronic mail. Electronic mail, a general term which is used to describe the transmission of messages by the use of computing systems and telecommunication facilities. It eliminates most of the problems and delays of getting a physical document from one person to another.

Electronic Media:

Electronic media is a media such as radio and television.

Electronic Shopping (e-Shopping):

Buying products and services online is called electronic shopping.

Electronic Marketing (e-Marketing):

e- marketing involves the use of the internet to market products and services.

Expert System:

Expert system is a decision support system for farmers and extension agents to obtain expert opinion and advice on field specific problems. It is a software application that attempts to reproduce the performance of one or more human experts.

Facsimile (FAX):

Facsimile involves the transmission of texts, pictures, maps, graphs, etc. between geographically separated points. An image is scanned at a transmitting point and duplicated at a receiving point.

Farm Advisory System:

These are services provided to assist farmers, firstly, to identify and implement improvements in farm management performance with regard to statutory regarding public, animal and plant health, environmental and animal welfare standards and secondly, to maintain their land in good condition.

File Sharing:

File sharing is an activity of exchanging information electronically, especially through the Internet.

File Transfer Protocol (FTP):

File Transfer Protocol support is a method of supporting remote networks. It is a protocol, which allows simple file transfer or documents.

Filtering:

Filtering is a process whereby a manager selectively accepts much inputs, which his mental ability can manage to process.

Firewall:

Firewall is a software that prevents unauthorized access to a computer, website, etc.

Gazette:

Gazette is a local newspaper or one that is associated with an organization.

Geographic Information System (GIS):

GIS is a systematically designed, spatially indexed approach for organizing information about places or regions in order to facilitate analysis of relationships between different social, economic and environmental variables.

Global Village:

A world that is increasingly interconnected by communication technologies and that is tending towards a global culture is global village. The internet and the World Wide Web are examples of these interconnections.

Graphic Design:

Graphic design is a craft of combining text and illustrations in design.

Graphic Format:

A graphic format is a canonical representation for problems of certain class.

Hacker:

Hacker is an individual who uses his or her computer expertise to gain unauthorized access to a computer system or a site either to learn about the system or to examine or manipulate its data.

Hard Copy:

Hard copy is a printed or filmed output from a computer device in a readable form.

Hardware:

The physical component of a computer system comprising electronic, magnetic and mechanical devices is called hardware. It is the computer's actual machinery and equipments, including memory, cabling, power supply, peripheral devices and circuit boards.

High Definition Television (HDTV):

High definition television is a television system that presents a picture that is wider than conventional television screens and has twice as many lines of scanning for increased clarity and detail.

http:

The abbreviation of http is hypertext transfer protocol. It is the protocol used for exchanging files on the world wide web. Web browsers are http clients that send file requests to web servers, which in turn handle the requests via an http service.

Hyperlink:

A word, symbol, image or other form in a hypertext document that 'links' or directly transfers the user, to another element in the document or to another site. The hyperlink is activated with a mouse click.

Hypermedia:

Hypermedia is the information retrieval system used for accessing texts, audio, video, etc., on the world wide web. A hypermedia navigation might include links to music, opera, harmony, composers and musicology.

Hypertext:

A system of storing texts, images and other files that allows for links to related text, images, etc.

Information Management:

It is the task of controlling information and its flow within an organization or system.

Information Retrieval:

A process of using or manipulating a database in order to extract some specific information from it.

Information System:

A information system is a set of interrelated components, with a clearly defined boundary, working together to achieve a common set of objectives.

Information Technology:

It refers to the resources used by an organization to manage information that it needs to carry out its mission. Information technology consists of computers, computer networks, other pieces of hardware, software that facilitates the system's abilities to manage information in a way that is useful for managers. Information technology may also refers to the computer based technology and telecommunications.

Informatization:

It is the process through which communication technologies are used as strategies for furthering socio-economic development.

Interactive Video Text:

The video text which is transmitted via telephone, cables, data line or private network is called interactive video text.

Interactive Video:

It refers to a video programme with which an individual can interact. It allows the user to take part in the sequencing and selection of video material and control the programme according to one's own pace of learning.

Interactivity:

Interactivity is the degree to which the participants in a communication process have control over, and can exchange roles in their mutual discourse.

Internet:

The internet is a global network of computers. It is highly interactive in nature and has the ability to readily allow each and every user to be a consumer as well as a provider of information that make the internet unique, as a tool of information technology.

Laptop:

Laptop is a computer set up in a thin briefcase, which is easy to carry and use, which is convenient for storing information and pictures.

Local Area Network (LAN):

A digital communication system capable of interconnecting a large number of computers, terminals and other peripheral devices within a limited geographical area.

Media-Wiki:

Media-Wiki is a free software open source wiki package written in PHP, originally for use on Wikipedia.

Memory:

Memory is a device or medium that can accept data, hold them, and deliver them on demand at a later time.

Menu:

Menu is a list of options from which a programme user can select any one in order to perform a desired action.

Modem:

An electronic device which connects a computer to a telephone line and converts between digital data in the computer and analogue signals on the telephone line is called modem.

Multimedia:

Multimedia is a computer system that allows the user to manipulate and use different types of media, such as text, sound, video, graphics and animation.

Network:

Network is an inter-connection of computer systems or peripheral devices with carriers and data communication devices for the purpose of exchanging data and information.

Networking:

It is the building up of an informal group of people to contact as a way to learn about job opportunities especially by means of Internet communications.

Podcast:

A podcast is a type of digital media consisting of an episodic series of audio radio, video, PDF, or ePub files subscribed to and downloaded through web syndication or streamed online to a computer or mobile device.

Polycom Videoconferencing:

Polycom is a popular vendor for videoconference products, provides hardware and software solutions for these purposes. Polycom units have traditionally been used for videoconference with large group of people.

Programme:

Programme is a set of sequenced instructions used to direct and control the operations of the computer in order to solve a problem or to perform a particular task.

Search Engine:

Search engines are powerful tools as they do the searching based on the information given. Search engines search information on the World Wide Web.

Soft Copy:

Soft copy is the computer output which is displayed on the screen of a terminal without a permanent copy.

Software:

The set of computer programmes, procedures and associated documentation related to the effective operation of a computer system is termed as software.

Teleconferencing:

A system in which persons sitting before appropriate screens see and talk to each other via a computer communication network is called telecommunication.

Teletext:

Teletext is a generic word for text based information systems available on a public or network basis. The best known is broadcast teletext.

Telnet:

Telnet was one of the first Internet protocol. Telnet is used to act as a remote terminal to an Internet host.

Terminal:

Terminal is an input/output device which allows a user to communicate directly with a computer system.

Transponder:

Transponder is a device mounted on a communication satellite which receives, amplifies and retransmits signals from earth stations.

Video Display Unit (VDU):

It is an input/output device that consists of a television like screen for displaying outputs and a keyboard for entering inputs.

Video Text:

Video text refers to an information system with which text and graphic information is transmitted and then received either by a videotext television or on an ordinary television set with an adapter unit.

Videotext System:

It is the general term used to describe personal computing/ communication networks that permit interaction between people and stored databases.

Web Browser:

It is a software programme, which facilitates to access the information and presents it on the screen and helps in navigation on the Internet.

Web Page:

A web page is a single unit of information called a hypertext document. A web page may consist of multimedia content such as text, images, sound and videos.

Webcasts:

A webcast is a method of broadcasting information via the Web. Webcasts differ from videoconferences in that the communication is one-way. A webcast is a media presentation distributed over the Internet using streaming media technology to distribute a single content source to many simultaneous listeners/viewers.

Wide Area Network (WAN):

Wide area network is a digital communication system which interconnects different sites, computer installations and user terminals, and may also enable LANs to communicate with each other. This type of networks may be developed to operate nationwide or worldwide.

Word Press:

Word press is a free publishing software and open source blogging tool and a content management system (CMS) based on PHP and MySQL which runs on a web hosting service.

World Wide Web (www):

World wide web is a wide area, hypermedia information retrieval initiative aiming to give universal access to a large universe of documents. It is also called as the Web, www, and w3, one of the most popular parts of the internet. The web is made up of hypertext documents stored on servers around the world. Hypertext documents contain links one can click on to move to another section of the page, to different documents, or to another type of internet resources.

CHAPTER 9

Management and Entrepreneurship

Entrepreneurship has been described as the "capacity and willingness to develop, organize and manage a business venture along with any of its risks in order to make a profit". The people who create these businesses are called entrepreneurs. Entrepreneurship is a process and entrepreneur is a person.

Achievement Motivation:

Achievement motivation is the expressed desire to do something well for its own sake rather than to gain power or love or recognition.

Administration:

Administration is the guidance, leadership and control of the efforts of a group of individuals towards some common goal.

Advertising:

Advertising is communicating with and influencing someone to do something. It is usually to buy a product or service and often to think something.

Agripreneurship:

Agripreneurship is defined as generally sustainable, community oriented, directly marketed agriculture.

Aspiration Level:

The standards set by individuals for themselves regarding the level they want to achieve with future performance is called aspiration level.

Attitude towards Credit Institution:

The degree of favourable or unfavourable attitude towards the credit institution from which the farmer obtains loan.

Audit:

Audit simply means official examination of accounts.

Authoritarian Leadership:

It is a form of giving leadership in which the leader(s) fix the goals and behaviour code of a group.

Authority:

Authority in an organization is the power in a position to exercise discretion in making decisions affecting others. It is the formal vestment of power in a specific status/position within the organization.

Break Even Analysis:

Break even point is the level of production at which the unit neither earns a profit nor incurs loss. Break even analysis aims at finding the behaviour of costs with volume of production, the production/sales level at which the firm breaks even, the sensitivity of profit to variations in cost, price and volumes, etc.

Budgeting:

Budgeting means stating the objectives both in numerical and financial terms. It indicates the allocation of funds for different sectors of the programme and is essentially a part of the planning process.

Bureaucracy:

It is a form of organization characterized by central leadership, hierarchical regulation of functionaries, decision process according to general rules and routine procedures, impersonal relations between functionaries and the general public.

Career or Development Training:

This type of training is designed to upgrade the knowledge, skills and ability of employees to help them assume greater responsibility in higher positions. The training may lead to the acquisition of higher degree or diploma by the employees, and motivate them to move up in the higher levels of administrative hierarchy.

Cash Flow Statement:

Cash flow estimates are prepared to ensure that the unit will have necessary cash with it and it will not face liquidity problem. The cash flow estimates include sources of funds and their dispositions (uses).

Committee:

Committee is a small group of knowledgeable persons to whom some matter is referred to mainly for detailed examination and decision.

Contingency Planning:

Contingency planning is planning for possible future environments which are not expected to occur but which may occur; if this possible future is widely different from that premised, alternative premises and plans are required. Contingency planning is done to solve an unforeseen problem, which calls for immediate action.

Coordination:

Coordination means establishing harmonious relationship between the efforts of individuals and groups for the accomplishment of enterprise objectives. It deals with synchronizing and unifying actions of a group of people.

Creativity:

Creativity is the ability to bring something new into existence.

Debentures:

These are long term instruments which are protected by a charge on the properties of the issuing company. Debenture is an acknowledgement of debt, given under the seal of the company and containing a contract for the repayment of principal sum at a specified date and for the payment of interest at fixed rebate per cent until the principal sum is to the company as security of the loan.

Debt Service Coverage Ratio (DSCR):

Financial institutions which provide the bulk of long term finance for industrial projects evaluate the financial viability primarily from the measures of DSCR and Internal Rate of Return.

Decision Making:

Decision making is the process of consciously choosing courses of action from available alternatives and integrating them for the purpose of achieving the desired goal.

Decision Support System (DSS):

A decision support system is an interactive computer system that is easily accessible to, and operated by, people who are not computer specialists, who use the DSS to help them plan and make decisions.

Delegation of Authority:

Delegation of authority is the vesting of decision making power in a subordinate by a superior which involves the determination of results expected from persons in position, the assignment of tasks to persons in a position, the delegation of authority for accomplishment of these tasks and the holding of people in position responsible for their accomplishment of these tasks.

Direction:

Direction is one of the essential management principles of extension. It is the way extension manager issues instruction to the people which indicates what is to be done.

Drone Entrepreneurs:

These are characterized by a refusal to adopt opportunities to make changes in production formulae even at the cost of severely reduced return relative to others. Such entrepreneurs may even suffer from losses but they are not ready to make changes in their existing production methods.

Economic Motivation:

It is the values or attitudes which attach greater importance of profit maximization as ends and means.

Empire Builders:

They are the persons engaged in creating chain of new ventures having an ownership. They have the qualities of vast vision, flair of innovation and managerial capability to build an empire for themselves.

Entrepreneur:

Entrepreneur is an innovator who invariably links up the innovation with the market to earn a profit. Entrepreneur is one who always searches for changes, responds to it and exploits it as an opportunity.

Entrepreneurship:

Entrepreneurship is a package of personality characteristics of entrepreneurs. It is a purposeful activity indulged in initiating, promoting and maintaining economic activities for the production and distribution of wealth.

Entrepreneurship Development Programme:

It may be defined as an action plan to create an entrepreneur with achievement motivation, who can develop business, can take strategic decisions, can cope successfully with the internal and external environment and can bear the risk of investment.

Evaluation:

Evaluation is the process of determining the extent to which objectives have been attained.

Fabian Entrepreneurs:

Fabian entrepreneurs are characterized by great caution and skepticism in experimenting any change in their enterprises. They imitates only when it becomes perfectly clear that failure to do so would result in a loss of the relative position in the enterprise.

Goal:

Goal is a distance which one is expected to reach during a given period of time.

Grievance Handling:

Grievance is an expression of discontent or dissatisfaction with any aspect of the organization. It may be real or imaginary, legitimate or ridiculous, stated or unvoiced, written or oral.

Imitative Entrepreneurs:

These are characterized by readiness to adopt successful innovations inaugurated by innovating entrepreneurs. Imitative entrepreneurs do not innovate the changes themselves, they only imitate techniques and technology innovated by others.

Incentives:

Incentives are direct or indirect payments of cash or in kind that are given to an individual or a system in order to encourage behavioural change.

Induction or Orientation Training:

Induction training is given immediately after employment to introduce the new entrant to the job. This helps in molding the individuals as per requirements of the organization.

Influence:

Influence is the informal way of getting things done.

Innovative Entrepreneurs:

An innovative entrepreneur is one who introduces new goods, inaugurates new method of production, discovers new market and reorganizes the enterprise. They are the persons who create an organization and remain engaged in their pursuits of innovation and creation of novel products and technology.

Innovative Orientation:

Innovative orientation is the degree to which an individual is relatively earlier to adopt new ideas than the other members of his social system.

Innovativeness:

Innovativeness refers to the creation of new products, market product, market combinations, method of production and organization and the like that enable the enterprise gain competitive advantage in the market.

Inservice Training:

Inservice training is the process of staff development for the purpose of improving the performance of an incumbent holding a position with assigned job responsibilities.

Intermediate Range Plan:

This type of plan is framed to carry out and coordinate the functions of organization. Such plans are detailed documents and usually framed for 5 years.

Internal Rate of Return (IRR):

The internal rate of return of a project is the discount rate which makes net present value equal to zero.

Leading:

Leading is the process of influencing people so that they will strive willingly and enthusiastically towards attainment of the organization and group goals.

Line Function:

Line function is one in which a superior exercises authority directly over a subordinate.

Maintenance or Refresher Training:

This training is offered to update and maintain the specialized subject-matter knowledge of the incumbents. This type of training is given to the employees to keep them at their peak performance level and also to preview them from getting into a rut.

Management By Objectives (MBO):

MBO is a system of management, whereby every department and every individual has defined objectives so as to accomplish results in key areas of organization during a specified period.

Management Information System (MIS):

MIS is a formal method of making available to management the accurate and timely information necessary to facilitate the decision making process and enable the organization's planning, control and operational functions to be carried out effectively.

Management Orientation:

The degree to which a farmer is oriented towards scientific farm management comprising planning, production and marketing functions of the farm.

Management:

Management is the process of planning, organizing, leading and controlling the efforts of organizational members and the use of other organizational resources in order to achieve the stated organizational goals. Management is the art of guiding the activities of a group of people towards the achievement of a common goal. It is the executive function of planning, organizing, coordination, directing, controlling and supervising an organization.

Managerial Entrepreneurs:

They are the persons who prefer to continue in the same venture by transforming themselves to fit the changing demands.

Market Survey:

It is defined as an objective and systematic collection, recording, analysis and interpretation of existing or potential markets, marketing strategies and tactics and the interaction between markets, marketing methods and current area potential products or services.

Marketing Mix:

All the marketing activities can be divided into four elements known as the Marketing Mix - Product, Price, Place and Promotion. They are often called as 4P's of marketing.

Microfinance:

Microfinance is the providing of small loans that are repaid with short periods of time and are essentially used by low income individuals and households.

Mobile Entrepreneurs:

They are the persons who leave the ventures as soon as venture is created. The entrepreneurial activity ends as soon as the venture is created.

Monitoring:

Monitoring simply means to keep a watch on what is happening. Monitoring is generally done at the operational level and involves collection and analysis of information. It is a management technique in which extension agents collect data on the way an extension programme is implemented and the problems encountered by extension agents with this implementation.

Net Present Value (NPV):

The net present value is the sum of the present value of all the cash flows associated with the project. One may accept the project if NPV is positive and reject if NPV is negative.

On the Job Training:

This is periodical or ad hoc training in the job, and is generally provided by the superior officer or specialists to the subordinate field staff. This training is problem or technology oriented and may include formal presentation, informal discussion and opportunities to try out new skills and knowledge in the field.

Opportunity:

Opportunity can be defined as an attractive and excellent project idea which an entrepreneur searches for and accepts such idea as a basis for his/her investment decision.

Organization:

Organization is the process of identifying and grouping work to be performed, defining and delegating responsibility and authority and establishing relationships for enabling people to work harmoniously together in accomplishing the objectives. Organization deals with allocation of work among executives and other employees, and one very important part of this allocation is the assignment of planning duties, it is also concerned with allotment of responsibility for direction and control and of course with responsibility for actual operations.

Organizational Conflict:

Conflict is a part of organizational life and may occur within the individual, between individuals, between the individuals and the group, and between groups.

Organizing:

Organizing involves establishing an intentional structure of roles for people to fill in an organization.

Payback Period:

The payback period is the time required to recover the initial capital outlay on the project. The shorter the payback period, the more desirable the project will be.

Place (Distribution):

This means how and where the product will be made available to the customers.

Plan:

A plan is a trap laid to capture the future.

Planning:

Planning can be defined as deciding in advance what is to be done. Planning is designing a course of action to achieve ends. Planning is an essential element of any organization and it is everyday activity. Planning is decision making and involves selecting and integrating the courses of action that an organization will follow to attain its objectives.

Power:

Power is the capacity to control others.

Preservice Training:

It is a process through which individuals are made ready to enter a certain kind of professional job, as in agriculture, medicine or engineering.

Procedures:

Procedures are the sequence of steps and rules, followed in implementing plans developed in intermediate and short range plans.

Price:

Price is the monetary value of your product.

Product:

The product is anything that is offered to someone to satisfy a need or a want. It may be a physical object or service.

Progressive Values:

It is a positive attitude of the individual to change and to the adoption of new ideas.

Projected Balance Sheet:

Projected balance sheet is prepared on the basis of profitability statement and cash flow statements. The position of share capital, term loans, sundry creditors, bank borrowings, current assets, etc. is determined at the end of each year according to the movements shown in cash flow and profitability estimates.

Profit to Total Capitalization Ratio:

This is worked out by dividing profit (before taxes with long term debt interest added back) by total capitalization (i.e. equity plus long term debt). This measure the earning power of both owned and borrowed capital.

Promotion:

This is the last element of marketing mix. Promotion means marketing communication. It involves any activity which influences people to buy the product.

Public Administration:

Public administration is an art and science as applied to the affairs of the state.

Responsibility:

Responsibility is the assignment of work to a specific position.

Risk Orientation:

It is the degree to which a farmer is oriented towards risk and uncertainty and has the courage to face the problems in farming.

Scalar Principle:

The clearer the line of authority from the ultimate management position in an enterprise to every subordinate position, the clearer will be the responsibility for decision making and the more effective organization communication will be.

Short Range Plan:

Short range plans are the plans of one year duration (annual plans). They are to carry out actual work and are very detailed.

Socio-Economic Status:

It refers to the position of a farmer occupies in comparison to others with respect to caste, occupation, education, family type and farming size, type of house, material possession, farm power, farm implements and social participation.

Staff Function:

Staff function is one in which the staff is meant to assist the line personnel. It is advisory in nature. They think for the head of the organization/office and give necessary support and advice.

Staffing:

Staffing is the process of selecting, maintaining and developing personnel in position to fulfill the organizational objectives.

Strategic Plan:

Strategic plan determines the major objectives of plans confined to objectives and policies of organization. They are long run plans which provide general guidance.

Supervision:

Supervision is selecting the right person for each job; arousing in each person an interest in the work and teaching how to do it; measuring and rating performance to be sure that teaching has been fully effective; administering correction when this is found to be necessary and transferring to more suitable work or dismissing those for whom this proves ineffective; commending whatever praise is merited or rewarding for good work; and finally fitting each person harmoniously into the working group - all done fairly, patiently and tactfully, so that each person is caused to do the work skillfully, accurately, intelligently, enthusiastically and completely.

Supervision:

Supervision is the method of indirect control and provides for development of personnel. This is more effective than direct control.

Sustainable Agriculture:

Sustainable agriculture denotes a holistic, systems-oriented approach to farming that focuses on the interrelationships of social, economic and environmental processes.

System:

A system is an organised/complex whole and assemblence or combination of things or parts, forming a complex or unitary whole.

Term Loans:

It is also called as term financing. These are long term loans carrying fixed rate of interest and repayable in 5 to 10 years in equal semi-annual installments. They are provided by banks, State Financial

Institutions, State Development Corporations, All India Financial Institutions like IDBI, ICICI, IFCI, UPT, LIC and GIC.

Virtual Organization:

Virtual organization is a network of geographically distant employees who are linked by electronic communication. ACT: Advanced Centres of Training

CHAPTER 10

Abbrevations

AEA: Agro Ecosystem Analysis
AEO: Agricultural Extension Officer
AICRP: All India Coordinated Research Project
AMC: ATMA Management Committee
ARS: Agricultural Research Service
ASRB: Agricultural Scientist Recruitment Board
ATIC: Agricultural Technology Information Centre
ATMA: Agriculture Technology Management Agency
BA: Beneficiary Assessment
BDO: Block Development Officer
BTC: Block Level Technology Centre
BTT: Block Technology Team
CA: Commodity Association
CAD: Command Area Development
CADA: Command Area Development Agency
CARD: Computer-Aided Registration of Deeds
CD - ROM: Compact Disc - Read Only Memory
CDB: Community Development Block
CDP: Community Development Programme
CDP: Cattle Development Project
CGO: Central Government Organization
CPM: Critical Path Method
CSIR: Council of Scientific and Industrial Research
DARE: Department of Agricultural Research and Education
DASP: Diversified Agriculture Support Project
DEO: District Extension Officer
DOA: Department of Agriculture
DOE: Directorate of Extension

DPAP: Drought Prone Area Programme
DPO: District Planning Officer
DSS: Decision Support System
DTP: Desktop Publishing
DWCRA: Development of Women and Children in Rural Areas
EAS: Employment Assurance Scheme
EDP: Entrepreneurship Development Programme
EEI: Extension Education Institute
ELC: Experiential Learning Cycle
EOAH: Extension Officer in Animal Husbandry
ETC: Extension Training Centre
FAC: Farmer Advisory Committee
FCCC: Framework Convention on Climate Change
FIAC: Farm Information and Advisory Centre
FIG: Farmer Interest Group
FLD: Front Line Demonstration
FO: Farmer Organization
FPR: Farmers Participatory Research
FSA: Farming System Approach
FSRE: Farming Systems Research and Extension
FWP: Food for Work Programme
GIS: Geographic Information Systems
GLW: Group Level Workers
GMF: Grow More Food
GMFC: Grow More Food Campaign
HYVP: High Yielding Variety Programme
IADP: Intensive Agricultural District Programme
IARI: Indian Agricultural Research Institute
IBDI: Industrial Development Bank of India
ICA: International Cooperation Administration
ICAR: Indian Council of Agricultural Research
ICDP: Intensive Cattle Development Programme
ICDS: Integrated Child Development Services Programme

ICICI: Industrial Credit and Investment Corporation of India
ICT: Information Communication Technology
IDP: Innovation Decision Process
IFCI: Industrial Finance Corporation of India
IMM: Interactive Multimedia
IRDP: Integrated Rural Development Programme
ISOPOM: Integrated Scheme of Oilseeds, Pulses, Oil Palm and Maize
ITDA: Integrated Tribal Development Agency
ITK: Indigenous Technical Knowledge
IVLP: Institutional Village Linkage Programme
JRY: Jawahar Rozgar Yojana
KGK: Krishi Gyan Kendra
KVK: Krishi Vigyan Kendra
KVS: Key Village Scheme
LAN: Local Area Network
LLP: Lab to land Programme
MADE: Management Decision
MANAGE: National Institute of Agricultural Extension Management
MES: Multi-Agency Extension Service
MFAL: Marginal Farmers and Agricultural Labourers
MSY: Mahila Samridhi Yojana
NAEP: National Agricultural Extension Project
NAIP: National Agricultural Innovation Project
NATP: National Agricultural Technology Project
NCRSS: North Central Rural Sociological Society
ND: National Demonstration
NDDB: National Dairy Development Board
NES: National Extension Service
NFSM: National Food Security Mission
NGO: Non-Governmental Organization
NHM: National Horticulture Mission
NIC: National Informatics Centre
NIED: National Institute for Entrepreneurship Development

NODP: National Oilseeds Development Programme

NRC: National Research Centre

NRCWA: National Research Centre for Women in Agriculture

NREGA: National Rural Employment Guarantee Act

NREP: National Rural Employment Programme

NSIC: National Small Industries Corporation Limited

NTC: National Training Courses

NWDRPA: National Watershed Development Project for Rainfed Areas

NYC: National Youth Council

NYK: Nehru Yuva Kendra (Nehru Youth Centre)

OFT: On Farm Testing

ORP: Operational Research Project

PALM: Participatory Analysis and Learning Methods

PANR: Participatory Appraisal of Natural Resources

PAR: Participatory Action Research

PEM: Participatory Extension Management

PERT: Programme Evaluation and Review Technique

PLA: Participatory Learning and Action

PMRY: Prime Minister's Rozgar Yojana

PPTD: Pilot Project for Tribal Development

PRA: Participatory Rural Appraisal

PRAP: Participatory Rural Appraisal and Planning

PTD: Participatory Technology Development

PURA: Providing Urban Amenities in Rural Areas

PVO: Private Voluntary Organization

R&E: Research and Extension

RA: Rapid Appraisal

RAAKS: Rapid Appraisal of Agriculture Knowledge Systems

RAP: Rapid Assessment Procedures

RCA: Rapid Catchment Analysis

RKVY: Rashtriya Krishi Vikas Yojana

RMK: Rashtriya Mahila Kosh

RRA: Raid Rural Appraisal

RRS: Regional Research Station
SAMATI: State Agricultural Management Extension Training Institute
SATCOM: Satellite Communications Technology
SATNA: Simple Approach to Training Needs Assessment
SAU: State Agriculture University
SDEO: Sub-Divisional Extension Officer
SEO: Social Education Organizer
SFAC: Small Farmers Agri-Business Consortium
SFDA: Small Farmers Development Agency
SGSY: Swarnajayanthi Gram Swarozgar Yojana
SHG: Self Help Group
SIDBI: Small Industries Development Bank of India
SIDO: Small Industry Development Organization
SIET: Small Industries Extension Training
SME: Small and Medium Enterprise
SMS: Subject Matter Specialist
SREP: Strategic Research and Extension Plan
SSI: Small Scale Industries
STI: State Training Institute
T&V System: Training and Visit System
TAR: Technology Assessment and Refinement
TAT: Thematic Apperception Test
TEIA: Technology Evaluation and Impact Assessment
TFT: Training for Transformation
TMO: Technology Mission on Oilseeds
TREAD: Trade Related Entrepreneurship Assistance and Development
TRYSEM: Training of Rural Youth for Self employment
TTC: Trainers Training Centre
UNCED: UN Conference on Environment and Development
UNFCCC: United Nations Framework Convention on Climate Change
VEW: Village Extension Worker
VLW: Village Level Workers
ZARS: Zonal Agricultural Research Station
ZEO: Zonal Extension Officer

CHAPTER 11

References

Ban, V.D. and H.S. Hawkins. 2002. Agricultural Extension. J.S. Offset Publishers, New Delhi.

Barman, U., R.K. Talukdar, P. Mishra and S. Borua. 2011. Fundamentals of Agricultural Extension. Dr. Updesh Purohit for Agrobios (India), Jodhpur.

Bhushan, V. and D.R. Sachdeva. 2015. An Introduction to Sociology. Kitab Mahal Publishers, Allahabad.

Chitambar, J.B. 2016. Introductory Rural Sociology. New Age International (P) Ltd, New Delhi.

Chole, R.R., P.S. Kapse and P.R. Deshmukh. 2015. Entrepreneurship Development and Communication Skills. Scientific Publishers, India.

Dipak, D. and B. Jirli. 2011. A Handbook of Extension Education. Dr. Updesh Purohit for Agrobios (India), Jodhpur.

Dipak, D. and B. Jirli. 2013. A Handbook of Extension Education. Dr. Updesh Purohit for Agrobios (India), Jodhpur.

Govind, S., G. Tamilselvi and J. Meenambigai. 2013. Extension Education and Rural Development. Dr. Updesh Purohit for Agrobios (India), Jodhpur.

Gupta, D.D. 2016. Advances in Extension Education and Rural Development. Dr. Updesh Purohit for Agrobios (India), Jodhpur.

Murugan, P.P., A.J. Rani, M. Asokhan and S. Palaniswamy. 2012. Rural Sociology and Educational Psychology as applied to Agricultural Extension. Thannambikkai Publications, Coimbatore.

Ray, G.L. 2011. Extension Communication and Management. Kalyani Publishers, Ludiana.

Sandhu, A.S. 2015. Extension Programme Planning. Oxford & IBH Publishing Company Pvt. Ltd., New Delhi.

Singh, A.K. 2012. Agricultural Extension: Impact and Assessment. Dr. Updesh Purohit for Agrobios (India), Jodhpur.

Singh, A.K. 2014. Agriculture Extension and Mass Communication. Dr. Updesh Purohit for Agrobios (India), Jodhpur.

Singh, A.K., L. Singh and R. R. Burman. 2012. Dimensions of

Agricultural Extension. Aman Publishing House, Meerut - 250002.

Singh, A.K. 2016. Agricultural Extension and Farm Journalism. Dr. Updesh Purohit for Agrobios (India), Jodhpur.

Supe, S.V. 1987. An Introduction to Extension Education. Oxford and IBH Publishing Co. Pvt. Ltd, New Delhi.

Sutherland, R.L., J.L. Woodward and M.A. Maxwell. 1974. Introductory Sociology. Oxford and IBH Publishing Co, New Delhi.

Vasanthakumar, J., H. Philip, R.K. Theodore and M.S. Nataraju. 2017. New Dimensions and Approaches in Extension Pluralism for Rural Development. Dr. Updesh Purohit for Agrobios (India), Jodhpur.